WHEN IT HITS THE FAN

WHEN IT HITS THE FAN

A HANDBOOK TO SURVIVING ANYTHING

LARRY MULLINS

HOBBLE CREEK PRESS
AN IMPRINT OF CEDAR FORT, INC.
SPRINGVILLE, UTAH

© 2015 Larry Mullins
All rights reserved.

No part of this book may be reproduced in any form whatsoever, whether by graphic, visual, electronic, film, microfilm, tape recording, or any other means, without prior written permission of the publisher, except in the case of brief passages embodied in critical reviews and articles.

The opinions and views expressed herein belong solely to the author and do not necessarily represent the opinions or views of Cedar Fort, Inc. Permission for the use of sources, graphics, and photos is also solely the responsibility of the author.

ISBN 13: 978-1-4621-1498-6

Published by Hobble Creek Press, an imprint of Cedar Fort, Inc.,
2373 W. 700 S., Springville, UT 84663
Distributed by Cedar Fort, Inc., www.cedarfort.com

LIBRARY OF CONGRESS CATALOGING-IN-PUBLICATION DATA
Mullins, Larry, 1936- author.
  When it hits the fan / Larry Mullins.
     pages cm
  Includes bibliographical references and index.
  Summary: Discusses what life would be like during a natural disaster, a government crash, or any other unexpected turn of events. Includes preparedness guidelines.
  ISBN 978-1-4621-1498-6 (alk. paper)
  1. Emergency management--Handbooks, manuals, etc. 2. Survival--Handbooks, manuals, etc. 3. Home economics--Handbooks, manuals, etc. I. Title.

HV551.2.M86 2014
363.34--dc 3

                    2014021189

Cover design by Angela Baxter
Cover design © 2015 by Lyle Mortimer
Edited and typeset by Eileen Leavitt

Printed in the United States of America

10 9 8 7 6 5 4 3 2 1

Printed on acid-free paper

I dedicate this book to my wife, Patsy.
Without her support, it would have never been finished.

Contents

ACKNOWLEDGMENTS IX

INTRODUCTION 1

1 TEOTWAWKI: THE END OF THE WORLD AS WE KNOW IT . . . 3

2 FOOD: THE TRIGGER 15

3 THE POWER GRID. 21

4 THE ENEMY 29

5 BUGGING OUT: GETTING OUT OF DODGE. 31

6 BUGGING IN: AT HOME 37

7 GOING UNDERGROUND 45

8 THE SURVIVOR 49

9 YOUR CASTLE 57

10 HOME SECURITY PLAN. 65

11 HOME DEFENSE WEAPONRY. 75

12 FRESH PLANT FOOD. 93

13 FRESH ANIMAL FOOD 111

14 FRESH FISH FOOD 127

15 ODDS, ENDS, AND SEAFOOD 151

16 PRIMITIVE SURVIVAL SKILLS 171

EPILOGUE. 191

ABOUT THE AUTHOR. 197

Acknowledgments

Thanks to Daniel C. Mullins, son and editor. At the end of a long day supervising his many archaeologists and other office workers, he comes home to rest; but before he can sit back and relax, he spends his time going over this manuscript to make sure I have been using the queen's English properly.

Thanks to Hailey Mullins, my granddaughter and part-time secretary. When she comes home from a long day of class, laboring with a full load of college classes, she turns to my manuscript to see if the computer has been fighting with me that day and to do anything else that needs doing.

INTRODUCTION

Journalists and English majors often write books with little personal experience regarding the subject they are writing about. There is a big difference, however, between those who research and write about a subject and those who write a book based largely on their personal experience. Most readers prefer the latter and want to learn something that couldn't be learned from a Google search. You should ask whether the author has anything to tell you that he or she has personally experienced.

I am currently in my late seventies. For at least sixty years I have worked in survival and law enforcement. My undergraduate and graduate studies were in psychology, which has given me further insight into the mind of a survivor. This combined experience gives me a unique skill set, and this book is based on that experience. Research is also included where appropriate, but the book is not only the result of research. None of us have had to live through a total doomsday scenario—thank goodness—but a hands-on survival and law enforcement background provides a reasonable idea of what to expect in such an event.

This book is not intended to be a step-by-step how-to book. Rather, the intent of this book is to provide a better understanding of what to expect when "the end of the world as we know it" (TEOTWAWKI) happens, and, more important, what we can do to survive such an event. There are a growing number of people who are making preparations for a crisis event; however, often preppers haven't thought deeply about what that crisis will look like. Without a clear picture of what specifically to prepare for, preparations are not always correct; we must see clearly what to expect if we are to make correct preparations. This book will bring the chaos of TEOTWAWKI, also known as "when shit hits the fan" (SHTF), into sharper focus. In some cases, that will be through information. In others instances, a kind of fictional narrative may be used to bring understanding to what our adversaries might be thinking. I will give some ideas about defending ourselves and ways to stretch our stored food. I will provide information on primitive survival, which, should you find yourself in that kind of situation, will improve your odds.

Many of my examples will be specific to the United States because that is the place I am most familiar with. However, the principles of survival can apply to almost any situation in almost any place in the world.

The term *urban survival* is a way of talking about the need to survive where we are, rather than just getting lost in the wilderness. Urban survival assumes a catastrophic event big enough to involve us all—or TEOTWAWKI. Not the end of the world infinite, but rather the end as we know it.

1

TEOTWAWKI

THE END OF THE WORLD AS WE KNOW IT

> "A world is not made to last forever, much less a race; that is not [God's] way."[1]
> –C. S. Lewis

There has been a lot of recent discussion about the Maya and how their calendar indicated the date of the end of the world. According to their calendar, the world would come to an abrupt end at sunset on the winter solstice in 2012. Bruce Scofield, in his book *Signs of Time*, notes that "according to Mayan inscriptions, the last age began on August 12, 3114 BC, and it will end on December 21, 2012 AD." He goes on to say that "mankind is now approaching the end of the entire creation epic."[2] Although 2012 has now come and gone, it still makes sense to prepare for one of these events. The Mayan's prediction certainly got us thinking about what we would do to survive an event like the one they described—especially because they weren't the only ones to talk about TEOTWAWKI, or the end of the world as we know it.

The Bible's book of Revelation also tells us that the end is coming. This book discusses the Four Horsemen of the Apocalypse. John tells us that the first Horseman (the white horse) brings war, the second Horsemen (the red horse) takes peace from the earth, the third Horsemen (the black horse) brings judgment, and the last Horseman (the pale horse) is death, and hell follows with him. The pale horse brings actual judgment, including famine, war, and death. John also describes the events accompanying these visitors, including great earthquakes, the sun turning black, and the moon appearing blood-red, the stars falling to the earth, the heavens departing, and every mountain and island being moved.

Where might we see indications of the Horsemen? Here are a few places where we might see some of the destruction described in the book of Revelation:

ELECTROMAGNETIC PULSE (EMP)

This, in my opinion, is our biggest concern. A 2005 article in the *Washington Post* detailed a previously unknown threat to America. According to the *Post*, an electromagnetic pulse is "a serious risk" and deserves the time and attention of researchers.[3] A single missile, carrying a nuclear weapon, detonated at the appropriate altitude, would interface with the Earth's atmosphere, producing an electromagnetic pulse radiating down to the earth's surface at the speed of light. The pulse would knock out the nation's stressed power grid and other electric systems; this effect would likely last for years. The loss of power would have a cascading effect on all aspects of US society.[4] Electronic forms of "communication would be largely impossible. Lack of refrigeration would leave food rotting in warehouses." Transportation networks would also break down, exacerbating the decline in food supplies. "The inability to sanitize and distribute water would also threaten public health." Urban fires and wildfires, without water and machinery to fight them, would rage unchecked. All of the factors would result in a rapid "breakdown of social order."[5]

Survivors would find themselves transported back to the United States of the 1800s. CIA director Porter Goss testified before Congress about nuclear materials missing from storage sites in Russia that might have found their way into terrorist hands, including Iran and North Korea.[6] SCUD-type missiles can be purchased on the black market for about $100,000 apiece. A terrorist organization might have trouble putting a nuclear warhead on a land-based target with the SCUD, but it would be relatively easy to launch and detonate a nuclear-armed missile in the atmosphere. The launch could also occur from a freighter in international waters. Al-Qaida, for example, is believed to own about eighty such vessels. Four missiles fired simultaneously would be enough to totally wipe out the electrical grid of the whole nation. Few Americans can conceive of terrorists bringing our society to its knees by destroying our electric grid and everything that runs off it. But we have been warned.[7]

The United States is a treasure chest of wealth and resources; any nation wishing to conquer the US would not want to destroy those resources. For too long, we have thought our enemies' mission would be to totally destroy the US, but that would make no sense. When we think of other nations, their hope would be to conquer us, while at the same time, leaving our great resources intact. Nuclear warheads landing on our soil would not accomplish that goal. An EMP, however, would remove our government to a large extent as well as our people, but leave our great resources in place. This is a likely scenario, and we need to be concerned about it.

I don't believe anything covers the issue of our destruction in past predictions or prophecies as well as an EMP. The crash of the electrical grid is a likely starting point for the TEOTWAWKI scenario. A nuclear warhead going off on the ground will produce a large but limited amount of damage. The same warhead, however, detonating 285 miles above the country will, in one moment, remove all of the technology we have come to rely on. Such an explosion will create an atmospheric tsunami with inconceivable pressure—a tidal wave of X-rays and gamma rays. Such an explosion would take out the whole electrical grid. Every electrical circuit, microchip, circuit board, and computer would be instantly fried beyond repair. And, unlike a normal power outage, the EMP-caused outage would be long lasting. Every bank, hospital, school, water treatment plant, plane, train, and automobile would be inoperable. Agriculture is now big business and—like other facets of our lives—is driven by technology; farms couldn't easily return to horse plows and hand milking. Over the course of a couple of months, the nation's population would likely be reduced to a shadow of today's numbers.[8]

It should come as no surprise to any of us, that the US has the same capability to cause an EMP and, if attacked by another nation, we would likely get missiles in the air to retaliate (if possible). It could turn into a worldwide crisis, with electrical power removed from multiple nations or, in a worst-case scenario, from *every* nation. It is thought by some that four warheads detonated high in the atmosphere could crash grids worldwide. This is one of the most likely scenarios for bringing about the end of the world as we have known it.[9]

SOLAR STORMS

From what we can see, our sun is a peaceful, benign object. But in truth, when it is seen through our powerful telescopes with the proper lenses, it is anything but. Our sun is in a state of constant activity. There are huge eruptions every minute that throw particles out into space for millions of miles. Our earth would be burned to a cinder if it weren't for our magnetic shield. There is a limit, however, to how much energy the shield can withstand. Every once in a while, our sun gets worked up into what scientists call megastorms.[10] The sun has done this many times in the past, but people on earth did not understand what was happening, so they just called it something else and went on with their lives. The sun's megastorms will have devastating effects on our electrical-based products; it will be the ultimate EMP. In 1800, few people even understood what electricity was. Yet, in the last two hundred years we have gone from horse-and-buggy lifestyles to a world totally dependent upon it. And *totally* is the correct word. Everything that moves our world today is dependent upon electricity.

What will happen when our sun begins to really acting up again? First, it will destroy all of our satellites, which will be enough to shut down our computers, handheld or otherwise. But just a little bit stronger, and sun storms could completely crash our electric grid, which would mean no electricity at all. Everything would shut down. Experts estimate anywhere from five to ten years would be required to rebuild our electric grid. "So what?" you may be saying. "I've got a wind-powered generator." But scientists tell us that in a severe sun storm, even local power generators would get fried. Simply put, there is no greater crisis we could face because the effect of this event will be worldwide. Most other natural disasters affect only a limited area, but sun-caused shutdowns would be worldwide and would last for years.[11]

Oil wells and refineries require electricity to refine petroleum products. Gasoline and oil-dependent trucks and trains are required to move that product to gas stations. In other words, when the electric grid goes down, so does the transportation industry. We will all be walking. Without electricity and transportation, we will also soon be without food. After a short rush, all grocery stores will be empty—the shelves stripped clean. We have come to see the grocery store as the only place where food can be obtained. Cafés, restaurants, and short-order establishments require the transportation industry to bring them the food they sell. Science is very clear here; this is not a matter of whether but when. We became too dependent too quick with no backup plan. The picture we need to see is more than we've allowed ourselves to see: there will be no food, no vehicles, no jobs, no money, no hospitals or clinics, no peace or safety, no recreational activities, no communication, and so on. I realize that our ancestors got along fine without electricity or gas, but their world was not built around it. Survival will depend on how quickly we can adjust to their way of life.

WATER SYSTEMS

If we should lose our power grid, water treatment plants will cease to do their job, and sickness will result. Water is the most precious thing we need to survive, and water treatment is more important than any other thing. Without clean, drinkable water, the rest is moot. Look at so many large cities like Los Angeles, New York, and Chicago; their water supplies come from far away, and they are helpless if anything should go wrong. If there's any interruption in that supply, there will be riots in the streets. If the water treatment plants shut down, and most surely would if the power grid went down, we would have to resort to disinfecting our own water and run the risk of ingesting all kinds of unpleasant bacteria. An individual can only go for a few days without water. Cut us off from safe drinking water, and we won't need to worry about the other dangers that may follow an electrical shutdown. Here's something to ponder: If your house no longer received water from the city, how would you flush the toilet? How long would it take for your house to become unfit to live in?

CYBER ATTACKS

There are groups intent on causing mayhem that have focused on bringing our nation down through other means. Their weapon of choice is the computer. The CIA has good intelligence that the government of China and one or two other countries have the ability to hack into our infrastructure and find ways to shut our power grid.[12] In addition, the FBI has channeled a large part of their budget into fighting cybercrime, especially that of individuals and nations bent on causing trouble. And then there are individuals working on the inside who are hacking into our own system every day. The FBI have been clear about their fears about above-average high school–age hackers who, simply for the challenge and bragging rights, would love to hack in to our system.[13] Government officials tell us this is a real possibility, and they are spending millions to try to stay a step ahead of the game.

ASTEROIDS AND COMETS

Asteroids, comets, and other space debris can be lumped together in the general category of space rocks slamming into the earth. The more important questions will be how big are these rocks, and what are they made of? Our earth takes hits every day, but the rocks are usually too small to cause much damage, and other space rocks usually fall apart in the atmosphere before they hit the ground. Some space rocks are made up of heavy metal, however, and these tend to survive the atmosphere and strike the earth. Occasionally, during the long history of the earth, space rocks have been durable and big enough to make sizable impact craters and—it has been argued—kill much of earth's living organisms.[14] The amount of debris that gets blown up into the sky is also an issue. If the strike is large enough, the amount of debris can be enough to block out the sun's rays for years, dropping temperatures enough to create winter-like conditions.[15] Governments and individuals are concerned enough to spend billions of dollars developing the technology needed to defend earth from space rocks.

RING OF FIRE

The earth's continents as we now know them were not always separate continents. There was a time when there was just one great landmass. That landmass began to break up and eventually form the continents we know today. At the spot where each continent broke away, there is a space or crack, which we refer to as a tectonic plate. The edges of these plates runs all the way around the many continents, and this is where all seismic activity takes place. Something must give when these plates collide, and generally one plate will be forced under the other. The effects include earthquakes, volcanic activity, and resultant phenomena like tsunamis. These events can, in turn, affect the weather. Geological experts tell us that the earth has, in recent times, been in a relatively peaceful condition, though it now seems to be moving back into a more active phase.

Some volcanic events could be severe enough that they would wipe out all life on earth. For example, the Yellowstone Caldera in Yellowstone National Park has the potential to kill everything within hundreds of miles of Yellowstone Park. The huge cloud of volcanic ash will circle the earth and could shut out the sun for decades.[16] The caldera has a geothermal heart large enough to hold all of Los Angeles. It feeds into a massive back chamber only five miles beneath the surface and, deeper underground, the main pots are continually building pressure. The Yellowstone Caldera has erupted numerous times, including at least three major eruptions. These eruptions occur on a regular basis—as steady as the Old Faithful geyser itself—and generally occur once every 600,000 to 800,000 years. The last eruption occurred nearly 640,000 years ago, meaning that we could potentially live to see the next eruption.[17] Based on evidence from past eruptions, the next major eruption could cover most of the continent with ash.[18]

There are many different kinds of volcanoes. Some volcanoes form cones of volcanic rock, some form cinder cones, and some do not form cones at all. The lava from the latter volcanoes simply flows away from the source. These are the volcanoes you see in Hawaii. They are fairly easy to live with (as long as you stay out of the path of lava). Mount St. Helens, while a big volcano, was by historical standards only a moderate volcanic event. She destroyed a fair amount of land but killed only a small number of people. However, the eruption of Krakatoa in Indonesia in the late 1800s destroyed an entire island and all life for hundreds of miles around. Ash was blown into the sky, impeding agricultural production on a worldwide level.[19] Crops could not be grown for several years, and agriculture came to a grinding stop. Many people responded by turning back to hunting and localized subsistence living. Large numbers of farm animals ended up on the table as food. Now try to imagine modern society and how we would respond to such an event. Only a small percent of today's population even know how to hunt, and that number is growing smaller each generation.

Most of us get our food from our favorite supermarket. If agricultural production were to stop, we would be largely helpless. Imagine a man walking into his favorite market only to find it empty. There is nothing on the shelves. Only broken glass where people had fought over every last crumb. What would his reaction be? His response would probably be, "I'm as good as dead." Unlike survivors of the 1800s-era Krakatoa eruption, modern man, removed from subsistence living, would have nothing to fall back on.

HURRICANES AND TORNADOES

Recently, we were able to get a sense for how devastating hurricanes can be as we watched Hurricanes Katrina and Sandy unfold before our eyes. Hurricanes are devastating in a limited area, and, as terrible as those hurricanes were, most of our nation felt the effects only through the media. While it felt like the end of the world to those who experienced it, a hurricane is not a true TEOTWAWKI—at least not by itself. But add other crises happening at the same time, and the results could be very different.

While hurricanes devastate our coastal areas, tornadoes do their damage inland. It doesn't happen very often, but it is possible for both a tornado and hurricane to happen at the same time and in the same location. It has happened in the past. Think of the devastation that would deliver. So while tornadoes are not the most intense disaster that could occur, if they were combined with other serious weather events, the results could be catastrophic.

TSUNAMIS

Recent tsunamis have shown us just how devastating these waves can be. There is an island off the coast of Africa that has begun to break apart. An area covering half of the island has slipped more than twenty feet and will soon tear off and crash into the sea.[20] Scientists have been able to chart the trajectory off the wave that it will generate—and the eastern coast of the United States will be its main target. The wave could be more than one hundred feet high when it hits our coast. Take a minute to think about what that would mean. How many large coastal cities would it wipe out? How far inland will the wave go before something like the Appalachian Mountains stops it?

GLOBAL WARMING

I'm not going to spend time on why this is occurring or who is responsible. The important thing is that it is happening. We all know, for example, that the polar ice is melting at an alarming rate—much faster than anyone had thought possible. Melting polar ice creates problems for the world's climate. Polar ice is mostly freshwater, and as the salt is pressed out, being heavier, it sinks and is pushed away.[21] This is the trigger that begins the movement of water in what becomes a great conveyor belt—water moving through all the seas of the earth. Cool, heavy saltwater moves in the deep oceans toward the equator. A number of things, including seismic events, cause the cooler water to begin warming. As it warms, it rises. At some point when it has warmed sufficiently, the water will begin to flow back toward the polar regions. This conveyor belt is the world's radiator. It's one of the things that makes life possible on this planet. If you were to shut this system down for any reason, our climate would change so radically that much of the world would become unlivable.[22] The issue with polar ice melt is that the more the freshwater polar ice melts, the less salt there is to begin the conveyor process. The river, or core current, would come to a complete stop. We don't need to wait for that to happen, however, because just slowing it down will be bad enough to mess up climatic norms. Less saline oceans mean life for ocean creatures changes drastically too, possibly leading to a drop in the number of animals we use for food. In addition, melting ice causes the level of the sea to rise, inundating the lowest the coastal areas of the world. This is a problem with serious consequences. Imagine, for example, what Hurricane Sandy would have been like if the ocean had been a foot deeper. It would have been even more damaging and costly.

SEVERE DROUGHT

Global warming has another side effect: longer droughts. We may need to start considering what to do if crops start failing due to these prolonged periods without rain. Lack of rain affects crops, which also affects animal stock. What would you do if the fruits, vegetables, and meat started disappearing from the grocery stores? During periods of severe drought, subsurface aquifers are relied more heavily on. An aquifer is nothing but a pool of water underground, and it can be pumped dry, just like every other body of water on the surface. This would mean that wells would have to be dug deeper to reach the water.

Drought has likely been responsible for the destruction of many past civilizations and could potentially bring about the end of the world as we know it. Food—or the lack of it—may eventually turn the Four Horsemen of the Apocalypse loose.

DISEASE, PLAGUE, AND SICKNESS

We are all familiar with what was known as the black death or the plague of the Middle Ages, but we probably aren't aware that these disease are still out there in our world, waiting for a chance to return. Even in recent years, there have been a number of cases reported in the US.[23] In today's world, we have literally millions of neighbors traveling by air all over the world—meaning any plague would take only hours or days to spread. We all think the government, through the CDC, will keep that from happening. Yet the strongest warnings and concerns are coming from the government.[24] AIDS, too, is an ever-growing problem throughout the world, but we have become unconcerned because it's not in our faces every day. In fact, there are a great many other serious viruses that just need a small mutation to bring on another black death.[25] And there is little question that such an event would trigger a panic that would turn the masses into mobs.

OPPOSING IDEALISM OR AUTHORITARIAN POLICIES

There are many groups throughout the world that can't tolerate the idea that there are others who don't share their position. They are bent on the idea they must conquer all others, or at least convert them to their position. There have always been a few who felt it their destiny to rule the world. Egypt and the pharaohs worked at that. The Assyrians came next, and for a time they were successful. They were followed by the Babylonians and the Persians. Then came the Macedonians under Alexander. After them, the Romans. More recently, we witnessed two world wars, begun by those who saw themselves as leaders of the world. So it should be not come as a surprise that there are modern groups working to accomplish the same thing.

It should now be clear that there are a great many "Horsemen" racing toward us at top speed, trying to see who will get to us first. The only thing we can do is to make informed preparations to ensure we survive. The end of the world as we know it is not something that might happen; it is something that will happen. Only those who are prepared will survive. Let that be you and me. Let's begin to make all of the preparations necessary rather than waiting to see what happens. A great many individuals today want to believe that the world as we know it cannot end. But many great civilizations (Samarians, Phoenicians, Egyptians, Assyrians, Babylonians, Persians, and many others) thought the same thing. Alexander the Great believed he would create a world that would last forever. And yet with the passage of time, it's evident that nothing lasts forever. However, you can give yourself a fighting chance if you make yourself ready and prepare properly.

NOTES

1. C. S Lewis, *Out of the Silent Planet*, (MacMillan, 1943).
2. Bruce Scofield, *Signs of Time: An Introduction to Mesoamerican Astrology* (Amherst: One Reed Publications, 1997), 114.
3. Eugene Volokh, "Smart guns, electromagnetic pulse, and planning for unknown-probability dangers," *Washington Post*, May 23, 2014, http://www.washingtonpost.com/news/volokh-conspiracy/wp/2014/05/23/smart-guns-electromagnetic-pulse-and-planning-for-unknown-probability-dangers/.
4. Committee on Armed Services House of Representatives, "The Report of the Commission to Assess the Threat to the U.S. From Electromagnetic Pulse Attack," No. 108–37, Washington Government Printing Office, 2005, http://commdocs.house.gov/committees/security/has204000.000/has204000_0.htm.
5. Jon Kyl, "Unready For This Attack," *Washington Post*, April 16, 2005, http://www.washingtonpost.com/wp-dyn/articles/A57774-2005Apr15.html.
6. Ibid.
7. Ibid.
8. Chris Stewart, *The Great and Terrible Fury & Light* (Salt Lake City: Deseret Book, 2012).
9. Ibid.
10. Richard A Lovett, "What if Biggest Known Sun Storm Hit Today?" *National Geographic*, last modified March 8, 2012, http://news.nationalgeographic.com/news/2012/03/120308-solar-flare-storm-sun-space-weather-science-aurora/
11. Gar Smith, "Flare-up: How the Sun Could Put an End to Nuclear Power," *Earth Island Journal* 27, no. 1 (2012), http://www.earthisland.org/journal/index.php/eij/article/flare-up_how_the_sun_could_put_an_end_to_nuclear_power.
12. Adam Segal, "China and the Power Grid: Hacking and Getting Hacked," last modified December 3, 2014, http://blogs.cfr.org/cyber/2014/12/03/china-and-the-power-grid/.
13. Federal Bureau of Investigation, "Computer Intrusions," http://www.fbi.gov/about-us/investigate/cyber/computer-intrusions.
14. "Meteorites, Impacts, and Mass Extinction," last modified December 1, 2014, http://www.tulane.edu/~sanelson/

Natural_Disasters/impacts.htm.

15. Chris Oliver, "ATLAS Project Funded by NASA," *Nā Kilo Hōkū*, 46 (2013): 1, http://www2.ifa.hawaii.edu/newsletters/article.cfm?a=631&n=51.

16. Ker Than, "Huge Magma Pocket Lurks Beneath Yellowstone Supervolcano," last modified December 18, 2013, http://news.nationalgeographic.com/news/2013/12/131218-yellowstone-supervolcano-eruption-magma-reservoir/.

17. "Questions about Yellowstone Volcanic History," USGS, last modified July 6, 2012, http://volcanoes.usgs.gov/volcanoes/yellowstone/yellowstone_sub_page_54.html.

18. Robert Hull, *Welcome to Planet Earth-2050-Population Zero* (Bloomington, Indiana: Author House, 2011).

19. "The Eruption of Krakatoa, August 27, 1883," Australian Government Bureau of Meteorology, http://www.bom.gov.au/tsunami/history/1883.shtml.

20. Steve Connor, "Scientists Warn of Massive Tidal Wave From Canary Island Volcano," Rense.com, 2001. http://rense.com/general13/tidal.htm.

21. "All About Sea Ice," National Snow & Ice Data Center, http://nsidc.org/cryosphere/seaice/index.html.

22. "Questions and Answers about Global Warming and Abrupt Climate Change," World Watch Institute, last modified August 1 2014, http://www.worldwatch.org/node/3949.

23. Tia Ghose, "Bubonic Plague Still Kills Thousands," *Huffington Post*, last modified September 27, 2013, http://www.huffingtonpost.com/2013/09/27/bubonic-plague-kills-thousands_n_4005495.html.

24. Donald G. McNeil Jr., "Wary of Attack with Smallpox, U.S. Buys Up a Costly Drug," *New York Times*, last modified March 12, 2013, http://www.nytimes.com/2013/03/13/health/us-stockpiles-smallpox-drug-in-case-of-bioterror-attack.html?pagewanted=all.

25. Leon Clifford, "Ebola: The Nightmare Scenario of a New Black Death," Expect to Be Challenged, last modified October 11, 2014, http://leonclifford.com/2014/10/11/ebola-the-nightmare-scenario-of-a-new-black-death/.

2

Food

THE TRIGGER

> "During the hyperinflation in post-WWI Germany, what used to be a comfortable nest egg was suddenly the value of a postage stamp.... There is a famous photograph, however, of a German woman during this time period burning piles of tightly bound banknotes to keep warm."[1]
>
> –Congressman Ron Paul

There is a story of a man who had a dream in which he was presented with a cat and a bag of gold. In the dream, he was told he could choose either the cat or the bag of gold but not both. He chose the cat. When he was asked why he chose the cat over the bag of gold, he replied that he could eat the cat but he couldn't eat the gold.

If you found yourself in a situation where every supermarket was an empty shell and restaurants and warehouses were empty, how much good would a dollar bill be? A room full of money would have no value. Your jewelry would also be worthless. In fact, one of the few things of real value would be food. Paying for a year or two's supply of food might seem like something you can't afford at today's prices. But at the time when all of the diamonds, rubies, gold, and silver are worthless, food will quickly become the new gold.

In the first chapter, I presented several scenarios that could create a catastrophic impact on worldwide food supply, and I would propose that the disruption or disappearance of our food supply is inevitable and will the trigger worldwide chaos. The majority of the population is confident that the government will keep this from happening. There is a lot law enforcement can handle, but they won't be able to fix all the problems when things get completely out of control. Some of us might still remember the riots in Watts, where less than a tenth of the population brought an entire police force to its knees. It took the police and the National Guard six days to get the rioting under control. This riot was not about the loss of food; it was about perceived police brutality. Future riots over the lack of food will not be brought under control until food is restored and people have confidence that the supply is stable.

I have spent a full career as a law enforcement officer, and I can tell you that emergency responders rarely get there in time to handle the problem. They get there in time to take a report. The only one who might be able to handle the problem is you. Is there danger? Sure. If the problem is about loss of property, it is often best to just let go. But if the issue is about life and death, your actions may be different. The loss of food can be a life-and-death issue.

The crisis will begin with some panic-causing event. It will probably involve fear of food shortages of some kind. We have seen it with the approach of a hurricane. People rush to the store and stockpile everything they think they will need. Even those little events can empty store shelves in mere minutes. This is not a big problem so long as trucks come within a day or two to bring the items needed to restock the shelves. But if there are no trucks coming, the little event becomes a big problem. That's when you get mobs, riots, and general chaos.

Here is a simple but likely scenario. We will start off with an individual law enforcement officer, whom we'll call Officer Smith. In the first couple of days of the food crisis, he will be out in the field doing what officers do. In this case, he'll be trying to control the mobs in the supermarkets with his fellow officers. They will not be doing much good, however, because the sheer numbers will overwhelm them. People will be so concerned about getting their carts full of food that they won't be able to be controlled. There will be a lot of pushing and shoving and fighting over everything. By the second day, the crowds will switch to restaurants and warehouses. By the third day, people will know that the whole transportation system is shut down.

As more serious panic sets in, the more aggressive members of society will take to the neighborhoods, trying to get whatever they can by force. On this kind of day, each officer will be faced with a new decision. Every man is at home trying to protect himself and his family. Officer Smith says to himself, "Do I stay out here on the street trying to protect others while my family may be fighting for their lives at home, or do I park my squad car and go home to protect my family?" This choice is a no-brainer. He is going home. And so are all of his fellow officers. So are all of the dispatchers, administrators, politicians, government officials, and members of the National Guard. Who then will be out there protecting you and your family? No one will be out there. Only you. We will see the loss of food supplies as the catalyst for a life-and-death struggle in our streets and neighborhoods.[2]

Of course our bodies need food to survive, but even more than that, we are all (in a way) addicts. We are all addicted to food. Psychologists tell us there is no addiction as strong as our addiction to food.[3] When we are cut off or think we are not going to be able to get food, it becomes our single focus. We will do anything to get our hands on food. It is hard to understand this because most of us have not experienced going without food for more than a day or two. Even in the survival courses I taught, the students knew the class tests would be temporary situations. But to be cut off from food indefinitely, not knowing if it will ever be available again, is totally a different experience.

This might be a good time to look back at an article in 2008 by Anthony Faiola of the *Washington Post*. While the facts in this article are a few years old, reports suggest that things have gotten worse in every respect.

> The globe's worst food crisis in generations emerged as a blip on the big boards and computer screens of America's great grain exchanges.... In Chicago, Minneapolis and Kansas City, traders watched from the pits early last summer as wheat prices spiked amid mediocre harvests in the United States and Europe and signs of prolonged drought in Australia. Within a few weeks, the traders discerned an ominous snowball effect.... As prices rose, major grain producers, including Argentina and the Ukraine, battling inflation caused in part by soaring oil bills, were moving to bar exports on a large range of crops to control costs at home. It meant less supply on the world markets, even as global demand entered a fundamentally new phase. Already, corn prices had been climbing for months on the back of moving government subsidies like ethanol. Soybeans are facing pressure from surging demand in China. But as supply is in the pipeline of global trade shrank, prices for corn, soybeans, wheat, oats, rice and other grains began shooting through the roof.
>
> At the same time, food was becoming the new gold.... By Christmas, a global panic was building. With fewer places to turn,

and tempted by weaker dollars, nations staged a run on the American wheat harvest.

Foreign buyers who typically seek to purchase one or two months' supply of wheat at a time, suddenly began to stockpile. . . . Food riots began to erupt worldwide. . . . US mills [began] to jump into the fray with their own massive orders, fearing that there would there would soon be no wheat left at any price. . . . Jeff Voge, chairman of the Kansas City board of Trade [said], "We have never seen anything like this before. . . . Prices are going up more in one day than they have during entire years in the past."

The food price shock now roiling markets is destabilizing governments, igniting street riots and threatening to send a new wave of hunger rippling through the world's poorest nations. . . . Prices leapt 80 percent. . . . "This crisis could result in a cascade of others . . . and become multidimensional problem affecting economic growth, social progress and even political security around the world," UN Secretary General Ban Ki-moon said.[4]

When money gets tight, what is the hardest thing to cut out of your budget? We can eliminate nearly everything but food when we need to. What I'm asking you to do is imagine clearly and deeply what our world would look like if food were no longer available to us. Try to picture it step by step. Every grocery store in town is empty. Transportation has halted. You understand that there will be no one coming to fix the problem. The only food you will have from now on is what you already have in your house. When that is gone, you will be without food of any kind. Also, try to picture what will be happening as all of your neighbors face the same crisis in their homes. What will they do? If you spend some time thinking through this situation, you will begin to understand what you'll need to do to prepare yourself and your family to survive. If you're serious about this and try to see clearly what you can expect to happen, I think you will discover several key things you will need to address.

Without question, the first thing you'll see is the need to get enough food to last you and your family for as long as you honestly believe the crisis will last. If any of the scenarios we have discussed occur, it will take years, not weeks, to create a reliable system of food production. You need to get serious about food storage. Without food, you are quite simply not going to survive. Every other part of preparing yourself for such a crisis situation will revolve around storing food and storing the proper amount of food. If you put off making the proper preparation for that day, you will either cease to live, or you will turn into someone you won't like.

Clearly, we must have food to live. Even more important is having the right amount of proteins, fats, carbohydrates, and vitamins. This problem becomes more serious in times of stress. My rule on the subject is, "No source of nutrition can be rejected in a survival situation." There are a great many sources of nutrition, even if you presently reject them, including things like worms, grubs, snails, and snakes. When you find yourself in a survival situation, you simply cannot afford these kinds of food prejudices. Nothing edible can be rejected. The situation will have reduced your intake supply to a large degree. To keep up your strength, you will need to eat anything and everything you can get your hands on. And you will need to understand that everyone around you will be obsessed with the same need and will be doing anything they can to relieve their hunger.

Many of us have a safe in our house where we keep things we value. I suggest you go to that safe and take out everything. Look at each item one at a time, and determine what the real value of that item is in light of what we have been discussing. In a world turned upside down, what is there of real value in that safe? Food could become the new gold. It will be the only thing of real value. Food is the key. Food is the trigger.

NOTES

1. Ron Paul, "The Benefits of Sound Money," *Daily Reckoning*, last modified March 13, 2009, http://www.dailyreckoning.com.au/the-benefits-of-sound-money/2009/03/13/.

2. Laurence Gonzales, *Deep Survival: Who Lives, Who Dies, and Why* (New York: W.W. Norton & Company, 2004).

3. Mark Cheren, et al., "Physical Craving and Food Addiction," The Food Addiction Institute, 2009, http://foodaddictioninstitute.org/scientific-research/physical-craving-and-food-addiction-a-scientific-review/.

4. Anthony Faiola, "The New Economics of Hunger," *Washington Post*, last modified April 27, 2008, http://www.washingtonpost.com/wp-dyn/content/article/2008/04/26/AR2008042602041.html.

3 THE POWER GRID

"[The] whole nation depends on technology.
Stop the wheels for two days and you'd have riots.
No place is more than two meals from a revolution."[1]

–Larry Niven and Jerry Pournelle

There are a number of events that could set off days of rampage and turmoil. However, I believe the two most likely are the collapse of the electrical grid and the collapse of the oil industry. We have discussed at length the many ways the grid could be brought down by natural means, but it is also possible that people could be responsible for bringing it down.

One of the FBI's biggest concerns is a cyber attack on the grid.[2] Terrorists are busy working on ways to sabotage the grid because they know its fall would quickly bring the United States to its knees. With that in mind, you might wonder why power companies are left to fend for themselves. Our present system is tied to several key power hubs called transmission system operators, and the reduction of just one of these will leave much of our nation in the dark. Take out multiple power hubs, and the whole nation will be without power. What the terrorists hope for is a method to shut down electrical power so that it will stay off for a long time. It is not unrealistic to believe that other governments might also have their sights set on disrupting US electrical systems. Finally, potential terrorists from within put us at risk. Any gifted high school hacker might be able to get the job done. At any rate, we are a lot more vulnerable than any of us realize. We need to anticipate what our world would look like in the days following a crash of the grid.

The system that powers our lights, refrigerators, and air-cooling systems also provides us with gas, oil, diesel, and all other petroleum products. With a massive electric crash, the end of the petroleum industry will quickly follow.[3] This dual event will likely be the trigger of coast-to-coast chaos. One could look at it like a row of dominoes—the first domino to fall would be the electrical grid. When it tips, it strikes domino number two, the petroleum industry, which in turn strikes domino number three, the transportation industry. And that process would continue until the whole line of dominoes is down. Without a transportation industry, food supply as we know it ceases to exist.

Explaining the process is important, so in order to make it easier to visualize, I'd like you to consider how a crash of the power grid might affect the typical American family. In this fictional account, we will call the father James.

One day around 6:00 p.m., the TV went dark. James tried for several minutes to get it to come back on. His sixteen-year-old daughter, Sandy, came into the room and complained her laptop had lost Internet access and she couldn't get it to come back on. She said that while she was trying to get the Internet back on, her battery had run out, and her charger wasn't working. James started to her bedroom to check on the problem when he noticed the little light that showed on the kitchen stove was off. He tried the kitchen lights, only to find they wouldn't come on either. Sure enough, the refrigerator was off too.

"It looks like the power is out," he said. Sandy sighed and headed back to her bedroom. James began to notice that the house seemed to be cooler than normal. He went to the control thermostat on the wall to turn the heat up. The blower would not come on. It took a moment for it to dawn on him that without electricity, the furnace wouldn't work either.

James's wife, Mary, had been taking a nap, and now she came into the hall to ask him what he was doing. He explained to her there seemed to be a blackout. She told him to call the phone power company and see if it was just their house or the whole neighborhood. It was then that he realized that not only his cell phone, but also the landline was completely dead. About that time, Jake, his fourteen-year-old son, came into the room and announced that his laptop was on but he couldn't get it to do anything; it wouldn't respond at all.

"So how are we going to find out what's happening?" asked James.

"I don't know. I went next door to ask, and no one seems to know. They don't have power either. It looks like it's out all over the neighborhood."

Mary's biggest concern was food; she had been sick all week and had not been able to do her weekly grocery shopping. She was complaining her refrigerator and cabinet shelves were getting empty. Since it was getting close to bedtime, everyone decided to just turn in early, sure the power would be restored by morning.

"Tomorrow is Saturday," said James. "We can go shopping in the morning and get all that we need."

"I can't go," responded Mary. "I'm still sick. You'll have to go. I'll make a list."

When they went to bed, they had to pile on more covers because the house was getting colder and colder.

The next morning, the house was so cold that James didn't want to get up out of the warmth of the blankets, and he slept in for an extra hour. Once up, he tried to push aside a nagging worry about the pipes freezing, but there was little he could do about it, so he tried to get his mind on other things. After a bowl of cereal and warm milk, he thought about the shopping he was supposed to do. He woke Mary up to ask for the shopping list she had promised. She sat up in bed and wrote down everything they would need.

"Looks like you're not planning to go with me?" James asked. She only groaned and pulled the blankets up to her chin. James realized he was on his own. He walked out to the garage and pushed the button to raise the garage door, but it didn't move. James then remembered: with no electricity, there was no garage door opener. He had to use the hand crank. As he backed the car into the street, he noticed less activity than usual. His favorite supermarket was only five miles from his home.

He headed in that direction but noticed the fuel gauge was getting close to empty, so he decided to fill up the gas tank before shopping. As he neared the gas station, he saw two long lines starting at the pumps and continuing back a full block before rounding the corner and going out of sight. The attendant at the station seemed to be pumping gas by hand, and a large sign painted on a piece of plywood said, "Cash only." Since he only had ten dollars in his wallet, he decided to wait until later when the lines would be shorter. He continued on toward the supermarket.

When he pulled into this store parking lot, he was met with a surprise. Not only was the lot totally full, the streets surrounding the store were packed as well. He began driving in circles. Around and around he went to each of the lanes, looking for someone pulling out. After a full twenty minutes of this, he finally saw a man and a woman loading the back of a minivan. He stopped and waited for them to finish. Many other cars were doing same thing, and three other cars had positioned themselves to get the spot before James. But when the couple finished loading the van, they slammed down the hatch and pulled their cart back into the store, presumably for more stuff. James drove on. Ten minutes later, he happened on a car leaving. Just as it pulled out, he swerved in front of an older couple to get into the spot before they did. He felt guilty about doing that, but he was getting concerned about the amount of fuel he might have left. As he started toward the store, he saw a shopping cart jammed in between two cars. With a crowd like this, he feared this might be the only cart he would get, so he dragged the wedged cart out, scratching one of the cars as he did.

He looked around and, seeing no one, headed for the door. When he got there, he found the real logjam, with more people trying to get in than people trying to get out. James positioned himself behind a huge man. He figured that would be the best way to get in quickly. It turned out to be a good call. The big guy pushed his way through the crowd with James in his wake. Once inside, James steered toward the meat section. The store was overcrowded. It was jammed so tight that one could hardly make any headway. The butcher was nowhere in sight, but there was nothing in the glass case anyway. The open shelves were almost totally gutted, and several people were fighting over the few things left. James could see this aisle would be a waste of his time, so he moved on.

The dairy section was empty too except for a bag of cheese that had a tear in the top. For some reason he couldn't explain, he took it. On the next row, he found two boxes of quinoa. He put that in his cart as well. He put a box of millet in his cart just because it was there. On the next row, he found two cans of something called jack mackerel, and a little further on he found ten cans of sardines. He now realized that he was cleaning out the last of everything. As he neared the front of the store, he spotted a fifty-pound bag of whole-wheat flour, looking like someone had put it in their cart and then changed their mind on their way out.

James moved to get in line at the cashiers, only to see there were no lines and no cashiers. Everyone was pushing through the store with no intention of paying. James had been planning to put everything on a credit card, but with no cashiers that wouldn't be possible. He thought about just leaving his cart, but somehow that seemed like a bad idea. He decided that when things returned to normal he would come back and pay for his stuff. He saw the big man pushing his way back through the crowd and he fell in behind him, as before, and moved out of the store. As soon as he was outside, he saw another big problem. All of the cars were trying to exit at the same time. With his short supply of gas, he had little choice but to wait until the lot cleared. Even with the traffic pouring out of all four exits, it seemed to take a long time for the majority clear the lot, since more cars just kept coming.

On his way home, James drove to the gas station, hoping things had improved. But what he found was the lines were longer than before, so he opted to wait a little longer. He drove home, and when he pulled into his driveway, he saw the garage door was down again. Grumbling, he put his car into park and, leaving the car idling, walked to the front door and into the garage to crank up the door. He then pulled the car in and packed everything into the kitchen. James turned to the boys' room. His son Jason seemed to be asleep, but Jake was sitting up doing homework.

"Jake, did you close the garage door?" he asked.

"I did. I didn't want anyone to rip us off while you were away."

James gave him a sloppy salute and turn toward the master bedroom. While Mary was awake, it looked like she had only just awakened, and she was pale and looked miserable.

James turned and headed back to the kitchen. He stood in front of the counter and spread things out. Mary appeared in the doorway.

"What have you got in here?" she snapped.

"What I've got are the very last items there were in the store."

"Well what do you want me to do with them? I don't even know what half of this stuff is, and I sure don't know how to cook it. You should have just left it there where you found it." She wrapped her housecoat tighter around herself and said, "When you are done with that, I need you to call and get me an appointment with the doctor for a strep screen. My throat is killing me."

"Listen carefully to me, Mary. This stuff on the counter, along with whatever you've got in the fridge and in the cupboards, is all we have to live on. There is no phone, and there is not going to be any doctor's visit today. Do you understand how serious this is?"

"Don't be silly, James. They will have the store restocked as soon as the power comes back on.

"You're not getting it. The stuff we need to restock the shelves needs to be brought in by truck. Let me tell you about my experience at the gas station."

James then went on to explain to her how long the gas lines were and about the fact that there didn't seem to be any gas coming. He explained to her that he had seen no trucks on the road and not many cars. He explained to her that gasoline was running out, so diesel would be running out. That meant there would be no trucks on the road and likely no trains.

"Nothing can be restocked without trucks and trains, and until then, there will be no food available anywhere, at any price."

"They will fix things," snapped Mary. James hoped she was right, but just in case, he checked the garage to make sure the doors were locked.

There are, of course, a great many ways to bring about an end of the world as we have come to know it. There is one thing most doomsday preppers seem to agree on—at some point, the power grid will crash. If the power grid were to crash, the first day would go something like this little story suggests. The first thing most of us would notice would be the sudden loss of electrical power. Since most of us have experienced one or more short-term blackouts, we would likely not be too anxious at first. It is also possible that without power, we might not be able to learn the full extent of the problem. It also likely that rumor, at least, would begin to get around in the first twenty-four hours. As the rumors start to spread, people would begin to panic.

We know from past experience, such as our experiences with hurricanes, what people's reactions would be. The first thing people will do is race to the store and stockpile essentials. Stockpiling food always comes first, and past experience tells us a food market can be emptied in one day. Next, people will make a run on gasoline and then hardware. Batteries, ropes, lamps, candles, generators, and things for the generators to power—like heaters, cook stoves, water pumps, and a host of other things—will go quickly. Then people will try to get everything they might need from the drugstore. They would gas up their vehicles and fill all of their gas cans. And this would be before they even knew for sure how serious the problem was going to become. People who wait until they get more information would be too late.

These are the only things that would affect individual homes initially, but having the power grid crash would have a profound effect on so many other things. People will be stunned as they begin to feel the impact of these new and perhaps unexpected consequences. For example, the whole petroleum industry will falter without electricity, in turn affecting transportation. At the drill site, at the refineries, and even at the pump, the whole industry will come to a stop without it. Without electricity, hospitals can't operate. Schools will close. Water treatment plants won't operate; drinking water has to have electricity to the city to pump it to your home. Even your worksite will shut down because it requires electricity to power computers and lights. Banks will close and lock their doors, because not only will computers be down, but everyone will also be trying to get their money out of the bank at the same time. In short, in this day and age, almost everything requires electricity. We have become slaves to electrical power. I fear we will not come to realize just how much influence electricity has over us until it's gone. If that power loss is extended, meaning it will be gone more than a few days, there will be other serious ramifications.

I am convinced that it is not a matter of *if* but a matter of *when*. When the grid crashes, the world as we have come to know it will change. Our whole world will be thrown back into the horse-and-buggy era. In addition to struggling to survive, we will have to relearn the skills of our forefathers. During the first year of the end of the world, whole nations will be readjusting. Population levels will drop as more and more people die. There is little we can do about that. The one thing we can do is to be among those who survive. You can's simply hope. You'll need to start preparing now. You must make all of the necessary preparations to ensure you have the best chance of survival. Then you will need to know how to stay alive during the period of time when the world is crashing into chaos. And lastly, you will have to prepare for life after the crash. I realize this is not a pleasant message. However, it is one that we all must have firmly planted in our heads if we want to live.

NOTES

1. Larry Niven and Jerry Pournelle, *Lucifer's Hammer* (New York: Tor, 1985), 44.
2. Federal Bureau of Investigation, "Cyber Crime," http://www.fbi.gov/about-us/investigate/cyber.
3. Jason Makanski, *Lights Out: The Electricity Crisis, the Global Economy, and What It Means to You* (Somerset, NJ: Wiley, 2007).

4 The Enemy

> "One of the common failings among honorable people is a failure to appreciate how thoroughly dishonorable some other people can be, and how dangerous it is to trust them."[1]
>
> –Thomas Sowell

The number one rule in any kind of conflict is know your enemy. This is not difficult in a war between two armies because both armies differentiate themselves from each other so their soldiers don't shoot their own kind. In a situation where the enemy could be anyone, however, determining who is the enemy becomes a larger problem.

In a situation where our world has collapsed—broken down to the point we would be facing death unless we can adjust to a whole new way of living, knowing our enemy is paramount to survival. In this kind of world, we must learn to recognize every enemy we might confront and anticipate their moves beforehand.

It's true that we may have to deal with criminal outliers, such as those people who manage to escape state prisons during a power grid failure and, for those in large cities, gang members. However, the number one enemy for us to deal with will be the people we have been living among all our lives—those who were our friends, neighbors, and coworkers—people who, prior to the collapse, were just good folks. They filled the seats in our stadiums when we went to the game. They filled the seats in the movie theaters when we went to the show. The big question is, what is that it turns those people into enemies? The answer is simple: desperation and lack of food. If you have enough food and water to see you through hard times, and they have nothing, then you will see another side of your neighbors. As soon as it becomes clear that there is no food to be had anywhere at any price, and that the only water available is badly contaminated, all the rules change. Now their only thought is finding someone who has food and water stored and to take it away from them. What are the lengths they will be willing to go to in order to take it away from you? The answer is they will do whatever it takes. If they are desperate enough, that may even mean killing you and your family. That would certainly qualify them as an enemy.

My twenty years as a survival expert and my work as a law enforcement officer convince me that our most significant concern with respect to protecting our families in the event of a natural disaster, a world war, a disruption in the power grid, or almost any other significant catastrophe, will be providing our families with the food, water, and shelter that they need to survive. Whether you are a doomsday prepper or just a single mom who wants to keep her family safe after an earthquake, your first concern will be ensuring that your family has access to a food supply. Unfortunately, most of us are so overwhelmed by the idea of how we will feed our families that we don't even take the first step. My goal is to make you think enough to start to take your own self-reliance seriously.

The key? A family who has stored sufficient food and water to see them through a year or more of crisis will have no reason to harm anyone, other than in defense of self and family. However, if you have failed to make that preparation, you will have no choice but to go out into your neighborhood and take whatever you need in any way you can. You will then become the enemy.

NOTES

1. Thomas Sowell, *Controversial Essays* (Stanford: Hoover Institution Press, 2002).

BUGGING OUT
GETTING OUT OF DODGE

> "A retreat is a place you go to live, not to die. Setting up a retreat is, for the most part, practicing the art of the possible. It is a matter of wisely and shrewdly identifying what you have available and turning it into something usable."[1]
>
> –Ragnar Benson

As TEOTWAWKI unfolds, survivors will make a decision about what to do, and many will opt for heading into the hills. They reason that when living in an urban setting, there are a maximum number of people trying to live on an extremely limited amount of resources. Many will choose to bug out rather than attempt to hunker down in place and fight over resources. However, the biggest problem with bugging out is in understanding when it's time to leave. If you are a little late on the getaway, you'll find the roads closed with stalled and out-of-gas cars. The highway will likely be nothing but a big parking lot. To successfully bug out, you have got to hit the road before others or have a relatively short path out of town.

You will also need to think about a strange contradiction in your bug-out plan. You are leaving the crowded city because of the huge population; you're planning to go into the woods or other areas where the population is low. But you are not alone. Many people are planning to head where there are no people. The remote woods might become where the largest number of people are. Let's say you have your wonderful bug-out place. It has everything a person could ask for. It is a beautiful meadow, way back in the deep woods. It has a good population of deer to offer you fresh meat and a nice little stream where fishing has always been good in the past. Then, on the day the SHTF, you'll rush your family into your bug-out vehicle and drive your family to your bug-out spot, only to find your meadow filled with tents, motor homes, pickups, and campers. Your lake and its stream are lined with people fishing. People are in your woods hunting. The city has moved to the woods, and the once-empty area is now filled with others. So what can you do to avoid this scenario? How you go about locating your bug-out home will be critical. If it is easy for you to get there, it will be easy for others to get there as well. Making the bug-out scenario work will require serious planning and research. This is not something you do in a hasty, willy-nilly way. You must invest as much (maybe more) time and effort into this as you did in planning the construction of your personal home. There are many serious considerations in developing your plan.

The first thing to consider is location. Where will you go? Consider issues such as population, climate, terrain, resources, and land ownership. You will need to know whether the land is available for you to occupy or if someone else already owns it. You may not think this is important in the end-of-the-world scenario, but it might be crucial. If you own the land, you have a better chance of controlling issues such as access and resources available on your piece of land. Because fresh water is the single most important thing to your survival, you should not consider a semiarid or desert location unless there is a dependable water source on your property.

Another thing to consider is the availability of dry wood. Few city dwellers understand how much wood they need to get through the winter. A reliable estimate is around ten to twelve cords. Not people even know what a cord is or how much work is involved in cutting and stacking it. One cord of wood measures four feet by four feet by eight feet of stacked logs. That's a lot of wood! If you're wondering why this is so important, remember that there will likely not be any electrical power or petroleum products available to you. Every bug-out plan must include a feasible option for getting by without either one. To best understand what you're going to need, start by understanding what you have the ability to survive without.

The next consideration is shelter. What will you live in? A tent or a camper? Perhaps a motor home or maybe a permanent dwelling like a cabin? The amount of time you are planning to live in it will help make the decision. The bug-out shelter will not be about a weekend getaway while waiting for someone to get the power back on. We are talking about TEOTWAWKI. You need to be thinking about living at your bug out locale long-term—perhaps for the rest of your life. For this reason, I believe a tent should be considered as a place to live in for only a short period of time, and that would also be true of a camper or motor home. In addition, your home back in the city will be standing alone and unprotected through the looting, riots, pillaging, and vandalism. It is possible you will not have a home to return to after a crisis. This is another compelling reason for you to build a bug out dwelling.

In clearing the site, you should be thinking about building materials. Materials could include logs, stone, earth, straw, or plaster. Perhaps a shelter could be shipped to your site, such as a geodesic dome or a yurt. Whatever you choose, it should be something you will be comfortable in for years. I spent most of one winter in a tepee and was very comfortable; however, I don't recommend that for the average family. One more thing to consider is how impervious your house should be. The looting and pillaging that will start in the city when a crisis starts will likely not stay in the city. There will be a time when the mob will decide to leave the city because no more food can be found. They will want to go out into the woods, where the people with food will be found. Your house should be built to withstand small arms fire. I will talk more about that in a later chapter.

You should select a place with freshwater and, with the water, a steady supply of fish. But again, things might not be like they are now. Resources like water will not be yours alone. You may be thinking that because you are a little better than the average fisherman, you'll do fine. But think about this: if peace officers are missing from the city, why would it to be any different in the country? Officers who enforce game laws will be taking care of their own families, not enforcing rules about fishing illegally or taking more than your limit. This scenario will deplete the population of fish overnight.

Most folks talk about hunting and plan to hunt deer, elk, and other animals to feed their families. Again, the problem begins with understanding your competition. With hordes of people out in the woods, killing everything available, the number of animals would be depleted in a hurry. There are a few ways to get around this problem that will be covered in later chapters.

Now let's deal with the important subject of bug-out vehicles. If you have selected a bug-out spot far enough away and hard enough to get to, then your choice is simple. The vehicle that takes you there must be something that can cover tough terrain. You must go somewhere others will not find or be able to navigate to. Simply stated, your bug-out place should be hidden and hard to reach. The road into your place should not look like a road. Ideally, it should look more like a trail or not be apparent at all—especially for the first mile. It should even be camouflaged. You cannot give passersby any choice or you will have a lot of unwanted company and competition for resources. This is one situation where you simply can't afford to share. What that means for you is that your vehicle must handle off-road conditions. A car simply will not do. Neither will a big motor home. My choice would be a good SUV or pickup with a long bed and four-wheel drive.

You will likely have to live in your bug-out cabin for a long time, so the next thing to consider is making yourself self-sufficient—which speaks directly to the issue of raising your own food. The first thing you will want to work on will be a garden. When you are deciding the location for your little hideout, you'll need to give some thought to the ingredients for a successful garden. During World War II, people were encouraged to grow a garden. The government told everyone that would make victory possible, and the small garden plots were called "victory gardens." Similarly, if you build a greenhouse, your odds of success will be better. You will not have the land needed for a herd of large-bodied animals such as cattle, but sheep and goats may not be out of the question. And with goats, you have a good supply of fresh milk. A chicken coop is a must for fresh eggs, meat, and fertilizer for your garden. Rabbits are also a good choice and are particularly good for meat. So are guinea pigs.

Though there are a number of ways to bug out, there is only thing that will determine the success of the effort: timing. You must be ahead of the crowd. You must be able to see the signs; you must be out on the highway before anyone else suspects there is a problem. There are bug-outers whose only plan is to get out of the city. They plan to pick up their bug-out backpacks and start walking. If they have the timing right, there should be no problem, because things are still proceeding naturally. But if they are even just a little out of time, it will be too late because there will be so many people trying to get out of the city.

Assuming you could make it out, what's next? It will do little good to have worked out the first stage of a six-stage plan if you have no idea what you're going to do with the other five stages. The majority of the bug-outers are planning to get into their bug-out vehicles and hit the road. Here, timing is even more important. They sense the time, feel the necessary urgency, see signs, and move out ahead of the others. The highways are still running normally. Getting out of town is no problem. If, however, you should you wait just a little for better confirmation, it will be too late. Before you can even get out on the freeway, traffic will have stopped. At first, you don't see this as a big deal. You've been here before. *Just an accident probably. They'll have everything moving soon.* But it turns out that isn't the problem, and they will not be moving again, maybe ever. What you cannot see is that the highway is one long parking lot, with everything just sitting there, idling until they start running out of fuel. After which they will not be moving again, ever. You can't move your vehicle, even if it's still running. You'll be jammed in. That is why timing is such a big issue.

You are left with two possible options. First, you can get out of your vehicle and start walking toward your bug-out spot. But if you have made a good choice about where it is located, you'll have many, many miles to walk. Maybe you've got a number of small children, making that option out of the question. Option number two is getting out of your vehicle and walking back to your house. As in every well-planned military mission, there is a fallback plan for when the primary plan falls apart. You, too, should have a fallback plan, or a plan B. There must always be something planned when plan A doesn't work out. In this case, plan B is what we do if we cannot reach our bug-out place. In reality, there are tons of things that could go wrong. Your plan A is likely to get out to your cabin, where you will have stored food, water, guns, ammunition, and everything else you will need to get through the hard times. But what if you simply cannot get there? Then what would you do? You would turn to your plan B. What is plan B? The only viable plan B is that you have already prepared your home in the city with all the same things. All of the food, water, guns, ammunition, and everything else you will need to succeed; in other words, you need to be prepared to be both a bug-outer and a stay-at-home survivalist.

How do you bring your family together in these plans? Let's begin with a husband and wife; nothing will work until they are in agreement about what to do in case of emergencies. They need to be on the same page, so to speak. With young children, there should be very little disagreement. They tend to go along with their parents. With them, it may be more useful to make it a game. It wouldn't be wise to tell them anything that would scare them or cause them to worry.

With the older children, however, it is entirely a different thing. Their views of the world have been largely influenced by friends, schools, and the media. These sources have told them things like guns are bad, the government will take care of them through sickness and old age, the national debt is not really a problem, there will always be plenty of law enforcement to handle the bad guys, the days of mob violence and race riots are in the past, serious natural catastrophes are rare, national infrastructure is well protected, and numerous other fallacies. Your task is to convince them that all of these ideas are not necessarily true. Not an easy task. Try to work around them. Get the united part of your family working toward your goal and hope a semi-united front might change the others.

The doubting individual may have more faith in others than they do in you. However, the upside of this (if there is an upside) is that you have prepared a plan. Then, if they refuse to go with you when it's time to bug out, at least there will be food and water in your house to keep them alive for a while. The downside is, if they are not willing or able to use a weapon to protect themselves, these resources will end up being taken and used by someone else. The doubters should be made aware what staying behind would mean. Tell them they need to know what will likely happen and know your bug-out plan is going to help you all survive. As you work on your bug-out plan, your children will be able to help. They will understand what life will be like once the crunch has happened; they will need to learn new skills such as self-defense, farming, fire building, home construction, and all of the many other things involved in survival living. Make it clear that you love them and that the whole family loves them and needs them. During this discussion, if they begin to appear less resistant, you can take that as agreement and then switch to giving them a list of responsibilities. It is important that every member of the family feels they have an important role to play and that each one sees that they are important to the survival of the family. Again, timing is critical both in the actual getaway and in getting your family's buy-in.

NOTES

1. Ragnar Benson, *The Survival Retreat: A Total Plan For Retreat Defense* (Boulder, CO: Paladin Press, 1983).

6 BUGGING IN
AT HOME

> "The cold reality remains that 'bugging-out' should be your absolute last resort.... Your chances of roughing it through a disaster at home are still much better than driving, biking, or walking out on your own ... possibly to a destination that is a worse situation than the one you are leaving."[1]
>
> –Sam Coffman

Television shows in the national media indicate that doomsday preppers tend to favor the bug-out option. When you wonder why, we quickly realize how people in general respond to the idea of worldwide disaster. It becomes clear that the first thoughts would naturally be to run away. It is a simple knee-jerk reaction, and unfortunately, most never process their plan for acting beyond thinking about it. That is the purpose of this book—to help you create a plan and practice that plan until you are comfortable with it.

Think, for example, of all the things that could go wrong in a bug-out event. Those choosing the bug-out option still need a plan B, or a fallback plan. So those favoring the bug-out plan would have to come up with enough money to pay for both plans A and B. Twice the food, twice the water, a separate bug-out vehicle, and twice the number of guns and ammunition. If you are a millionaire, you may be able to afford this plan, but many of us are working stiffs who live from paycheck to paycheck. By choosing the stay-at-home option, you will have to pay only once. That is worth thinking about. And I believe your chances of making it through TEOTWAWKI at home are at least as good, if not better, than driving, biking, or walking out with your family. You could find the situation in your bug-out place is worse than what it would have been at home. If bugging out is so popular, then much of the population will have chosen that option already, and you will find yourself living with many others. Again, timing is a critical issue when bugging out. If you wait to hit the road, even by minutes, you might not get out of the city at all.

The most likely cause of a large-scale collapse is the collapse of the power grid. Moving out because of the crash of the power grid suggests another reason why holing up at home makes sense. If the power grid crashes, things would not gradually move into a crisis scenario. A grid crash will be catastrophic enough to require overnight changes in the way people conduct their affairs. You would not have weeks or months to figure out what was happening, and without many of the forms of communication we are accustomed to having, it will be very difficult to get accurate information. Your response will need to be almost immediate. You probably can't afford to wait even a day to act. Here's what I envision as the scenario at the onset of a grid crash:

- Grocery stores would quickly empty.
- Gas stations would close down because there would be no product to sell.
- Hardware stores would empty.
- Water treatment plants would no longer be able to treat drinking water, and water would no longer be pumped into our homes.
- All businesses would close their doors, including your place of work.
- Hospitals would cease to function.
- Garbage would pile up all over the city.
- Sewage would back up and eventually spill over.

In this event, home will be the best place for you to put up a spirited defense. (That is, of course, unless you live in an apartment building. Then staying put might not be an option for you.) For this reason, redesigning your home into your castle—creating a location where you can make a stand against looters, rioters, pillagers, and worse will be your family's best defense. You can create a place where you can store food and water needed to get you through to better times ahead.

No matter which option you choose, the best way to success is to have all of your preparations in place before problems arise. If I were to choose the bug-out option, I would want to have my new outdoor home finished and stocked, and I would want to be already living in it. That's the only way I could be sure I would be able to enjoy the fruits of my labors.

Likewise, if I were to choose to stay put and ride it out in my own home, I would want to have all of my reconstruction in place before I needed it.

As far as duration, there is no way to know how long we will have to defend our homes. If my home is ready to stand against small arms fire, and if it has been set up so I can see everything moving my way and I can cover all of my positions from every side, then and only then could I prevail. Without question, we will be thrown back to conditions as they were a hundred years ago, and those conditions will last for an extended period. We will be living like our great-grandparents lived.

We have been told over and over again that our law enforcement system can handle the small percent of our society that is lawless. But that claim is built on the assumption that there is only a small percentage of people living lawlessly, so the claim is weak when it comes to a time of crisis. For an example of what we will deal with, we only have to look at the Watts Riots in California. The police force was fully manned and added plenty of officers to handle the so-called small-percent lawbreakers. According to the civil rights digital library,

> The riot spurred from an incident on August 11, 1965 when Marquette Frye, a young African American motorist, was pulled over and arrested by Lee W. Minikus, a white California Highway Patrolman, for suspicion of driving while intoxicated. As a crowd on onlookers gathered at the scene of Frye's arrest, strained tensions between police officers and the crowd erupted in a violent exchange. The outbreak of violence that followed Frye's arrest immediately touched off a large-scale riot centered in the commercial section of Watts, a deeply impoverished African American neighborhood in South Central Los Angeles.[2]

A small percentage of the population took place in the riots, yet it took nearly a week to bring them under control. It took the entire local police force, as well as officers from San Diego and the National Guard, before order was restored. In truth, the police force in any city is doing well if they can just handle the local criminals. Adding large-scale crime is more than they can hope to handle. Usually, you can count on one of the two Fs—frustration and fear—to understand how riots get started. The Watts Riots fall under frustration, but the kind of riot we are talking about is fear. There is little question that fear is a much stronger motive than frustration; if all of the police officers together with the National Guard couldn't handle Watts, how much hope would we have if there were a worldwide, fear-driven food riot?

The early part of this run on grocery stores would be slowed by employees trying to get money from the customers. This will be anything but easy with no power or cash registers working. The checkout stands would have to try to add up purchases and take cash because the banks would be closed and checks would be useless, as well as credit cards. You can imagine this would not last long with shoppers fighting to get all they could then get out of the store quickly so they could turn around and come back for more. It wouldn't take long before the employees would give up and get out of the way. All of this would be followed by a fear-driven run on gas stations, hardware, storage, and eating establishments. But that will only be the first stage.

Later, perhaps the next day or a day or two later, will turn to the next stage. As word gets out that stores are empty and there is no food to be had, a deep fear will get into people's hearts. "What can I do?" will be the first question. Over the course of that day, different answers will come to different people. Some will wait and see what will happen. Some will wait because they think that the government will fix things and get the power working again. Some will think they will count on friends and family until it dawns on them than everyone is being affected equally. Others may tell themselves they will be able to get by on hunting and fishing until someone in the family reminds them that on a good day of fishing they rarely came home with more than two or three fish. And even though the hunting season was several weekends long, most of the time they failed to fill their tag. It wouldn't take too long for people to realize that if there was any food to be had, it will come from the homes of people who had stored food. A couple more days of listening to their children cry because their empty bellies hurt will cause parents to begin to feel desperate. Then they will be looking for friends or family to help them go out and get food.

This brings us to the other side of the coin. That should be you—the person who has already fortified the home and stored sufficient food, water, and weapons to see things through the hard times. The only real issue left to work through is what you're willing to do to protect yourself, your wife, and your children. It won't matter whether you have chosen to bug out or stay put, one thing will be certain: people will come to take those things from you. Let's take a minute to be specific about what that means. If you encounter another desperate person who is intent on taking anything of value from you, you may have to deal with rape, torture, and death. These are people who will take your food, water, weapons, ammunition, and likely your life. Why would they feel they had to kill you? The answer is, "leave no witnesses." At this point in time, they will still be afraid order could be restored at some point, and they wouldn't want anyone left to point them out to the authorities. Please don't hope to negotiate. That may have served you well in the past, but in this world it will only make you an easy target.

Your first rule needs to be "never let them see you." Even if you know them or think you can trust them, do not expose yourself. Even when you think they are not a danger to you, you should still talk through a closed door. The world has changed. Nothing is the same. Unless you can be absolutely sure you know them and trust them, you cannot give them a chance to get inside your defenses. To retain the tactical advantage, you must keep them off your porch. And never, if there is even the slightest question, do you open your front door. That door is the most critical spot in your defenses. To open it is like dropping the bridge over the moat and throwing open the castle door. When your front door is breached, the battle is over.

You may be thinking that the bug-out option sounds safer, but that would not necessarily be true. If you are in your own home, there is little to do but defend. But with the bug-out option, a good part of the plan involves doing things outside. For example, you have to go outside to garden or to take care of chickens, rabbits, sheep, goats, and pigs. You will also have to leave to go hunting or fishing. When the pillagers arrive, you would most likely be outside dealing with some of those tasks, and you would not have the safety of your home. It wouldn't matter if you were a bug-outer in the deep woods; the attacks will come.

The medieval rule of war suggests that it would take a force nine times larger to have success over a good defensible position. In other words, you will have the tactical advantage over the attackers so long as you don't drop the drawbridge. At least initially, looting activities will move slowly. It will take a while before the looters realize what they have to do next. But when it begins in earnest, it will move more than fast enough. They will have no way of knowing where the food might be; they will most likely just move from house to house, taking whatever they can find. This will go faster than you might think, because few families in today's world have any way to resist. We have become such an anti-gun, anti-violence culture that most homes have little to defend themselves with. This might, however, be something of an advantage to those who *can* put up a spirited defense. A few well-placed gunshots might cause the looters to think the risk is too high at your house, causing them to just pick up and move on down the block to the next house. Cowards don't like to take chances.

Once the rioting gets fully underway, it won't take long until every neighborhood has been sacked. Then things will move to a really ugly stage. Many will say, "This could never happen," but I say it's time to study history. These types of full-scale riots have happened time and time again. The fall of Jerusalem to the Romans is a good example. The city had lost its ability to resist. The people had consumed every crumb of food. The strong began to prey on the weak. They were eating the dead, and worse, killing the weak to cannibalize them. Everyone left alive thought only of his or her own needs. I know you think that could never happen in America, but the issue is no longer influenced by where you live. Everyone will be trying to stay alive.

Looters, realizing they have exhausted all the easy pickings, are left with two possibilities. Some will hike out to the woods to find the folks who bugged out, perhaps thinking that they will still have food. Some of the others will remember the houses that resisted them, thinking they must have had something to protect. Why else would they resist so strongly? For the stay-at-home preppers, this is when the real fighting begins. Up to now, defense only required firing a couple of rounds. But now the pillagers will be returning to seriously test the defenses. Your plan has to be putting them all down; your shots must be center of mass. Don't try to stop short of deadly force. When it's over, don't try to move the dead. Their bodies are powerful statements about your ability to defend your home. It will go a long way in convincing others that you are not to be dealt with. Going out to move them will also put you in danger. You will be happy to learn that that stage three is temporary. The worst of the fighting will be over quickly.

In this chapter, I have talked a lot about the need to redesign your house into a place you can defend, yet I have said little about how that might be accomplished. Later chapters will deal with the specifics of creating your own fortress. In the bug-out approach to survival, the key is good timing. With the stay-at-home and dig-in approach, the key is preparing your home. Once you have a truly defensible home, you are more than halfway there.

NOTES

1. Sam Coffman, "Bugging In: Urban Survival at Home," *Survivalist Magazine*, no. 9 (2013): 40.
2. "Watts Riots," Civil Rights Digital Library, last modified November 20, 2013, accessed January 27, 2015, http://crdl.usg.edu/events/watts_riots/?Welcome.

7 GOING UNDERGROUND

"The power of population is indefinitely greater than the power in the earth to produce subsistence for man."[1]

–Thomas Malthus

In the 1940s, the whole country was obsessed with bomb shelters. We were all concerned about attack, and back then we were sure the USSR would bomb us any day.[2] The US and the USSR were like two pit bulls staring at each other across a fence, and building underground bomb shelters was a no-brainer. Finally, we negotiated an agreement to do away with the missiles. We shut down our silos; we even sent investigators to make sure everyone followed through. But that kind of thing is hard to regulate. Who knows how many might still be out there, waiting to be used.[3]

Let's not forget that Pakistan and India (both armed) have, in the recent past, been at each other's throats. India even issued a notification for residents to secure their basements for protection from an attack. Israel and Iran have also shown no qualms about trying to intimidate one another with weaponry, as well as several other countries with a score to settle.[4] Terrorists are clamoring to get their hands on some form of nuclear device to send our way. All in all, it would seem the need for bomb shelters is more serious today than in the past. Additionally, something like a storm shelter would be a good thing to have for those living in areas prone to natural disasters.

However, nuclear war or disasters are not the sole concerns of this chapter. Here we are going to look at the reasons for going underground as a survival strategy for TEOTWAWKI.

There are preppers who plan to use this strategy as their one and only way of overcoming hard times. Those who plan to go into their bunkers will try to stay put until the crisis is over. There are those who also plan to have them somewhere close to their home. Along this line of thinking, there is a film called *Taking Shelter* that won a great deal of praise at the Sundance Film Festival. I recommend watching it because it teaches a lot about building an underground bunker.[5] In addition, there are those who have built or plan to build their bunker up in some remote place in the woods. There are generally three different approaches to the whole subject of urban survival.

1. If the problem will be a short-term one, lasting anywhere from seventy-two hours to a couple of weeks, bugging-in may be required. A bunker might be a good idea.
2. If the problem will last a bit longer, perhaps six months to a year, bugging in or out for a while will certainly be required.
3. If the problem is on a TEOTWAWKI scale, you may want to look at using the bunker along with, or in addition to, a bigger, more livable dwelling.

Your planning will be a reflection on your point of view. Let's start with the stay-at-home, or bug-in, option.

In this option, you set up your defensible personal dwelling. Then you add your underground bunker somewhere close by to enable you to easily access it from somewhere in your house. That entrance must be well camouflaged. If it cannot be found, the bunker then becomes your ace in the hole—your last chance for survival. Your bunker should be fireproof and bulletproof. Depending on the situations you think you might encounter, you may also want to make sure it has enough strength to withstand bomb blasts.

In this first scenario, imagine having put up a good fight but realizing you're going to be overrun. Pillagers are breaking in, and you cannot hold out any longer. However, you still have your ace in the hole. You rush into the bunker and shut the door. You have already stored your food and water and everything else you will need to survive. If it's all in your bunker, there will be nothing left for the looters. When they discover the house is without food, they will have no reason to hang around looking for you.

Now let us consider the second scenario. You will certainly be bugging in or out. If you have found your bug-out spot and you have taken the time to build your dwelling, farm shed, garden, you will also need to take time to build a bunker. Adding a bunker system is vital. Build it in such a way that its entrance is inside your personal dwelling. That entrance, as stated before, should be a secret. Camouflage its entrance. If you are forced to use it, you do not want the enemy to find it. The bunker could be below your home with the access point placed in the floor of your basement.

Again, let's imagine what the situation might look like to incoming enemies. Somehow, if they manage to locate your place, then you can expect a larger group of pillagers. These are people who may have banded together with others as they moved away from the cities. There is greater possibility they will be able to overrun you. Once you and your family have moved into the bunker, you will be leaving pillagers an empty house without your food and water. Other things are already in the bunker with you. In this situation, the doorway into your bunker must be so well hidden that someone searching for it will not find it. They will take time to look for you. They know you didn't get out of the house, so they will suspect something.

One idea you might consider is a planned sacrifice. While you're in the planning process, you should accept the very real possibility that pillagers will find you. So you should plan for just such a thing, and you should not forget why you have prepared in the first place—so you and your family can stay alive. Nothing else should come close to that. So what if you can sacrifice your home to keep your family alive.

When I started planning for TEOTWAWKI, I devised a system where I would be able to set off a huge, fast-burning fire to the whole house if I could see we were about to be overcome. I designed a system where I could turn also turn all of my livestock out of the sheds. My family and I would be safe in our bunker. Our food and water would also be safe with us. And our livestock would at least have a chance to run into the woods. Sure, the looters would find some and slaughter many of them, but there would likely still be some left so we could start over after the crisis was past. The house burning down would say to the looters we chose to commit suicide rather than be captured. With nothing left but ashes, they would have little choice but to move on. The fact that we would all still be alive would be the payoff, and sacrificing the primary home would be a small price to pay. For us to come out and rebuild beats the alternative.

The form and nature of the actual bunker is up to you. I will say, however, that there are a number of builders who build wonderful underground bunkers, but they are expensive. Another good choice is to purchase a metal shipping container, the kind you would see on a cargo ship. These can be turned in and rebuilt into a bunker not unlike the ones you would purchase. While these are not cheap, they are a much better solution than anything you could build on your own.

NOTES

1. Thomas Malthus, "An Essay on the Principle of Population," 1798, https://www.marxists.org/reference/subject/economics/malthus/ch01.htm.
2. "Red Scare" History.com, http://www.history.com/topics/cold-war/red-scare.
3. Bill Text 113th Congress FY 2015 National Defense Authorization Act, Library of Congress, http://thomas.loc.gov/cgi-bin/query/z?c113:S.2410.PCS:/.
4. Gardiner Harris, "India Warns Kashmiris to Prepare for Nuclear War," *New York Times*, last modified January 22, 2013, http://www.nytimes.com/2013/01/23/world/asia/indian-officials-advise-preparations-for-possible-war.html.
5. Writer and director Jeff Nichols, "Take Shelter," released November 10, 2011, United States, Produced by Hydraulx Entertainment and Grove Hill Productions.

8   THE SURVIVOR

> "Most people can't think, most of the remainder won't think, the small fraction who do think mostly can't do it very well. The extremely tiny fraction who do think regularly, accurately, creatively, and without self-delusion—in the long run these are the only people who count."[1]
>
> –Robert A. Heinlein

In today's world, the average person is a hothouse plant. We have to go back several generations to the early 1800s, for example, before we find average citizens with the ability to live off of the land. It is important to know what TEOTWAWKI will look and feel like. But it's also critical to know what new attitudes we will need to survive. What are the things that make us able to survive?

When I was only three days old, I was turned over to my maternal grandparents, and they raised me. So I jumped over a whole generation in my upbringing. My grandfather became a father to me, and he was an old rancher with little patience for newfangled things. Instead of teaching me how to throw a baseball, he taught me the things he knew—the things he was good at. I tell my grandchildren that my grandpa was John Wayne long before Marion Morrison ever played the part in the movie. He was a superb horseman; successful rancher; and a highly skilled tracker, trapper, and fighter—he was a woodsman in the extreme. He taught me what plants were useful and which ones I could eat. He taught me how to find water and how to build shelters. He taught me how to build fire in many different ways without matches. He taught me to recognize edible mushrooms and poisonous ones. Most of the time, we were at the ranch or a line camp, so I rarely used electricity. Life was good—wonderful really, and I didn't realize things were different without electricity. Fresh meat had to be eaten quickly or it had to be cured. Most things don't keep well without a refrigerator. The funny thing was, I didn't know I was roughing it. In today's world, most youngsters are proud of their ability to handle video games. I was proud of my ability to shoe a horse, hitch up the wagon, or tie a good diamond hitch. There are few folks today who know how to make a rope.

As a young man, I went into the army and spent most of my time in the Panama Jungle Warfare Training Center. I managed to get most of my assignments out in the jungles because I was more comfortable there. My most common assignment was teaching jungle survival. When I returned home, I continued to do what I knew—teaching survival. I taught classes all the way from Alaska to Mexico. I have continued teaching survival for much of my life. It shouldn't surprise you, then, to know after teaching many thousands, I have learned something about what kind of attitude an individual needs if they're going to have hope of surviving.

Over the years, your body has become adjusted to a certain type of food, but given time, it can learn to do well on other types of foods. It will take time, and trying to do this during a high-stress survival situation is not a good time to ask your body to work on that change. For the first month, at least, your body will want what it is used to. This would not be a good time to try to become a vegetarian, for example. Your body will adjust better if you can maintain the same amount of animal protein, carbs, fats, and glucose. In the many years of running survival, I learned that a certain amount of sodium is also necessary. Animal protein is the hardest to obtain and the hardest to prepare, and it has been the hardest to get students to learn to live without. I have found most people in today's world are totally unprepared and unwilling to deal with a lack of meat. Most think that meat is something that comes in a cellophane package. While many people love their hamburgers, they are unwilling to deal with what it really is and where it came from. One statement I have to use on my students is, "Tonight we are going to be eating part of the dead cow." Now and then, someone catches on and says, "I wouldn't want to be eating a live one." On every survival experiment, I made sure my students had experience with killing and butchering an animal and cooking and eating it. Most of my survival trips lasted anywhere from two weeks to two months. Students would work together, removing the pelt and butchering the animal. There were two parts of the process—the first would be to cook and eat all the meat we could on the spot, and the second part would involve preparing the rest for the future by smoking and drying it. This was done by cutting meat into strips and hanging the strips over a smoking fire for a day. Making sense of reality is crucial to survival. Everyone must learn how killing and butchering is done not only to understand the process but also to prove to themselves that they can do it—that they can do the so-called dirty work.

You may find yourself wondering why. The simple answer is because others may not always be around to do the dirty work, and you will have to care for yourself. If one of the parents of the home was the only one who had ever killed an animal for food, wouldn't the family be in trouble if that person was gone? In fact, in a time of chaos, anyone might be taken out of the picture. That is why it is so important that every member of the family should have proven to themselves and be able to do whatever is necessary to keep the others supplied with fresh animal protein. Do not wait until the last hour. Now is the time for the family to have that kind of experience. Any time you eat a hamburger, you need to take responsibility for what you are doing. I've had students tell me they would not let themselves think about anything like that before the hamburger was served to them on their plates. That attitude is totally without responsibility.

We've all heard the expression "only the strong survive." There is a lot of truth in that. It's possible for others to survive, but only if they have someone strong and prepared to lead and direct them. When you're up against the big crash, you will be dealing with the greatest calamity ever known in the history of mankind. This will, quite simply, be a time of such turbulence, chaos, turmoil, and violence that the weak and unprepared will not have a chance. No one should be looking on this as a temporary inconvenience. To survive, you will need to get serious about the coming turmoil. This was clearly understood back in 500 BC in a quote often attributed to the Greek philosopher Heraclitus: "Out of every one hundred men, ten shouldn't be there, eighty are nothing but targets, and nine are the real fighters. We're lucky to have them. They make the battle, but one of them is a warrior. He will bring the others back."

There is no way to know how to teach anyone how to become a warrior, but I truly hope to provide some of the information you will need to be effective. I realize that every family is not alike; many of you may lead your children as a single parent. Some families will include grandmas and grandpas. Some may be in this as a combined family. I cannot address each of you separately, so I will write as though you were a family made up of a father, a mother, and several children.

In TEOTWAWKI scenario, the family that will have the best chance of surviving will be the family that pulls together. In that family, the husband and wife are a team. Let me tell you about my wife. We've been together longer than most of you have been alive, clearly illustrated by the ten children, thirty-nine grandchildren, twenty-six great grandchildren, and one great-great-grandchild. During all those years, we have walked side by side, pulling as a team. Neither one of us surpassed the other. And she is tough. Any adventure I took on, she was right there with me. If I were to be removed, I don't have a single doubt that she would take over and do the job every bit as well as I could. She can kill a rattlesnake and eat it as quickly as anything else needing to be done. A family that has a united mother and father will have an improved chance of survival in the coming crisis. I have pictured a family with a couple of teens, preteens, and small children. I understand that your family will be somewhat different. Just adjust my comments to fit your family. The more help you can get from your children, the better off you will be.

If you have teens in your family, they should be almost as valuable as adults. They can begin an adult assignment, like the defense of your home and family. The preteens can also be helpful. You may not be aware that among most American Indian tribes, boys as young as ten to twelve were already going on raids and learning to become warriors. In the Civil War, boys eleven and twelve years old were in uniform. They will be able to handle the stress better if they have something they see as important to do. Girls of the same age can be just as useful. As for your small children, some of them may not be as useful, but they still need to have some way to feel like they are involved. To park them off in the corner with nothing to do will only raise their stress levels higher than they need to be.

Ideally, the whole family needs to be operating as one. There is no time more important for that unity than when you come under attack. But that is not the time for trying to get organized. Everyone should have their tasks already drummed into their heads. That is the time for everyone to automatically go to his or her post. They need to go with full understanding of what they're supposed to do. This is not the time for the fighters to be distracted by the little ones. Giving the little ones something real to do will help ensure the others can stay focused.

A survivor's mindset also requires a conscious change in attitude. For example, some of the things available as food will be things you think you will not be able to eat. Most common responses go something like this, "When I am really hungry, I will eat that, but I couldn't do it right now." This is perhaps the biggest misunderstanding in the survival field. My personal experience is just the opposite. The truth is—the longer one goes without food, the harder it is to return to eating anything. I remember a time up in Alaska when I went seven days without food of any kind, and trying to make myself eat at the end of that time was difficult. If I had to eat something that didn't sound good to me before, it would've been impossible. As it was, I could keep nothing down but a few wild blueberries, and that was for the first day after my seven-day fast. You and your family need to learn now so you can eat later when you need to.

You also need to begin thinking about weaponry too, so you can move ahead and start to prepare yourself. This will not be a time with a lot of foolish idealism. You cannot think of yourself as a serious prepper while standing on an anti-gun soapbox. Things like, "I hate guns. I can't stand to be around them. You'll never get me to touch one," is nothing but foolish talk. How can you picture a dozen men standing on your front lawn, shooting out your windows, and trying to break down your front door and continue making such statements? Let me tell you what the model is with the real preppers: "I will fight until I'm dead. They will have to kill me to stop me. My family deserves nothing less." Ted Nugent said, "To my mind, it is wholly irresponsible to go into the world incapable of preventing violence, injury, crying, and death. How feeble is the mindset to accept defense this mess. How unnatural, how cheap, how cowardly, how pathetic."[2]

Having a defendable structure is absolutely necessary. It doesn't matter if you are a bug-out family or planning to stay put, you are still going to need a structure that can be defended. For a family planning to stay put, you might make a big mistake in thinking your home is already a sound structure that you can count on to keep you safe. It will need to be redesigned and reconstructed to make it defendable. The walls need to be able to stop, at the very least, small arms fire. The doors may need to be replaced. The windows need to be designed so that they can be counted on to stop bullets and, at the same time, be used as firing ports from which you can return fire. You need to develop good field of fire positions on every side of your house. This will be covered in a chapter totally devoted to the subject further on. As to your bug-out structure, you will be building it from scratch, and you will be able to incorporate all of this when you start.

The most important thing to build, though, is your attitude, and you will not survive unless your attitude is in the right place. The way you see the world must change as the world changes. The days of wine and roses will have disappeared, and the new world is going to be a very different place. You must be able to shift gears. You'll have to change if you hope to survive. Many folks simply ignore the possibility that something bad will ever strike them. Many believe that even making preparations for a disaster is likely to bring it on. Their position is not to talk about it because it will cause the thing to happen. That kind of thinking is not only defeatism, but it is also totally wrong. There is nothing as important as good planning. Accepting that disaster can happen is the first step, then making a good plan to handle whatever it might be. Every good plan should include a good backup plan or what to do if plan A doesn't cover the situation as well as you hoped.

Sam Coffman refers to the four A's, starting with:

> ATTITUDE: You cannot survive anything if your attitude is not where it needs to be. . . . Your mind is your greatest tool and greatest weapon. Attitude is the lubricant that allows everything to work correctly.
>
> ADAPTABILITY: You must learn to become adaptable because there is no way to predict everything that can happen.
>
> AWARENESS: Being aware of what is going on around you gives you an immense advantage in every aspect of survival. It also requires much less energy to avoid a bad situation than to try and get out of one.
>
> ACCOUNTABILITY: Being prepared for a disaster is not a lone-wolf activity. . . . If you want to function well as a team or community, you must be accountable for both good and bad decisions and be honest and sincere in your dealings with everyone you are working with.[3]

The serious survivor must develop a strong survival and combat mindset. This is what it will take to turn you into a real survivor. "It is one thing to say it, another to do it, and quite another thing to do it in the stress of mortal danger."[4]

NOTES

1. Robert A Heinlein, *Time Enough for Love* (New York: Ace Books, 1988).
2. Ted Nugent, *God, Guns & Rock'N'Roll* (Washington, DC: Renegery Publishing, 2000), 10.
3. Sam Coffman, "How to Prepare for American Armageddon," http://www.secretsofsurvival.com/survival/american-armageddon.html.
4. Laurence Gonzales, *Deep Survival: Who Lives, Who Dies, and Why* (New York: W.W. Northon & Company, 2004).

YOUR CASTLE

"In an apocalyptic disaster, you need more than just the door shut. Your home is most likely going to be your last line of defense."

–Unknown

It doesn't matter if we're talking about a bug-out or a bug-in structure, the key issue will be always be defending it from pillagers. But how we can hope to be safe against people attacking our house? If we construct our home correctly, to be a good defensive fortification, we should have the upper hand. The big question is how defensible is your home?

Everything you will be trying to protect or defend will be clustered inside the walls of your own home. And, while it may sound a bit chilling, the truth is, your home will be the obvious battleground where you will have to make your stand. When you imagine making a stand anywhere else, you realize you will lose that advantage. Your home is where you will be, where your family will be, and it is where you have your food stored. Your house is your stronghold. It is your castle. It is where someone will come looking for things of value.

We have clearly established that the pillagers will be seeking your food. This brings us back to the old story about the famous bank robber Willie Sutton; when he was asked why he robbed banks, he replied, "Because that's where the money is."[1] But money will no longer be the thing of value, and food will have become the new gold. Where will they be looking for food? Clearly not in the supermarket if the shelves are empty. The next logical place to look for food will be in people's homes. The Willie Suttons of the future will be looking to break into homes to find food.

There will be many different kinds of people coming to your home hoping to acquire food, and it will always be your choice as to how you want to respond. For example, your response will likely be very different toward a hungry-looking mother standing on your front porch with a baby in her arms and three young children huddled around her, while she begs for some assistance, versus four large men pounding on your door and demanding that it be opened. They are not there begging for a handout. They are announcing their intention to rob you. Now, I can't say what you should do in either case. But one thing should be very clear: you would feel different in your decision concerning these two different types of visitors.

If you are married with a family, your love for them is one of the main reasons for your stand to protect them. This is why protecting your food supply is imperative. None of those you care about and love will be able to survive without food. It is easy to say to yourself that food is just a material possession, but it is beyond that—it's a matter of life and death. As much as you think giving your food to hungry beggars is the good thing to do, it might mean hardship for your family later on.

There is a much-loved story about a family of ants who worked hard all summer to stockpile all the food they would need to get themselves through a long, hard winter. They didn't know how long the coming winter might be, so they simply worked as hard as they could to stockpile as much food as they could while the sun was shining and the plants were giving up their harvest. Throughout the long sunny summer, their friend the grasshopper simply lay around, enjoying the warmth of the sun. Then, almost overnight, winter came with great cold and a severe wind. The grasshopper began to search for food. It had been abundant just yesterday, but now he could find nothing. After a few days of that, the grasshopper began to panic. He was cold and hungry and getting weaker every day. Soon, he swallowed his pride and went to the ants, begging for a handout. But the old ant king, having lived through severe winters in the past, realized that if this winter were also hard, many of the ants would die. And so he turned his friend away. He was not being cruel; he was being fair. The grasshopper, by his own choice, wasted away the whole summer while the ants worked hard. It wouldn't be fair for many of them to have to die because the grasshopper was lazy and unwise. "You see," the old ant king told the lazy grasshopper, "we have just enough to see us through until the good times return."

In truth, we are in a situation like the ants, and anything we give away may come back to haunt us before good times return. You and I have been living in a relatively prosperous period. We haven't had the common experience of those in the past; we haven't had to go to bed each night wondering what kind of pillagers will come during the night to burn our homes, plunder our possessions, or murder our families.

The consequences of these golden years are seen in the current construction methods employed in our homes. They are no more defensible than cardboard boxes. With only a few exceptions, small arms fire could penetrate the walls of all but a few dwellings. Mobs, trespassers, and pillagers could gain access with very little effort. We pull our door shut, lock it, and immediately think we are safe. But what makes us feel so secure when a person could come along behind you and break open that same door with a reasonably quiet kick? Or, should an intruder decide to come in by some other direction, they could enter through any window in your house. In truth, you are not safe in your home at all. So it seems the first thing you need to do is try to make your home as strong possible. What kind of things should we change so we can have some confidence that we can defend ourselves, our families, and our stored food? What are the weaknesses in the structure of our home that can be strengthened, fortified, and improved?

It is important that you become confident in your own ability to protect your home because you will not have anyone but yourself to count on when the SHTF. Police officers will already be dead or in their own homes, protecting and defending their families. In law enforcement training, every officer is taught to use force equal to the force brought against him or her. An officer may not be forced to use a handgun, but it will always be there just in case that level of force is brought against him or her. You may not be forced to use a firearm to defend your house, but you had better have at least one gun available to you. I have a couple of friends who say, "I don't mind owning a gun, just in case. But I sure hope I never have to use it." Well, the only way I can think of making that dream come true is to make your house look like a castle—so solid and indestructible that it would need a small army to even start an assault. Then and only then can you hope not to have to resort to your gun.

Intruders will look over your castle to see if you have forgotten some part of your defenses. If they cannot find one, they will likely move on to the next house, looking for an easier target. Let us assume that you are not making preparations to hide, but making preparations for serious defense.

In this day, homes are not built to offer much in the way of protection. They are little more than cardboard boxes. The average house in America could not hold out a determined threat. Homes can be built to hold off determined threats, however, as illustrated in the Ned Christie story.[2] Ned was a Cherokee farmer living in what was then the Cherokee nation, now the northeastern part of Oklahoma. The story took place somewhere between 1872 and 1879. At this time, the Cherokee were allowed their own judicial court system and their own officers of the court. On the day that started Ned's problems, a trial was being held, and the Cherokee judge asked Ned to serve as a backup bailiff. During the trial, violence broke out, and the judge was killed. Word reached Judge Isaac Parker in Fort Smith that there was some question as to Ned's part in the shooting. Judge Parker set a posse to bring him in for questioning, but Judge Parker failed to make clear to the posse that Ned was just a person of interest, so they elected to bring him back belly down on his horse.

What they didn't know was that Ned was the best man in the Cherokee nation with a gun. The posse managed to burn down his house. After shooting him in the eye and leaving him in the burning house for dead, they returned to Judge Parker to report. The judge was not happy. Not only was his only good witness dead, but two-thirds of the posse lay dead around the Parker home. Later, when it was discovered that Ned was still alive, the white law insisted that he be arrested for killing a bunch of lawmen. Although he tried, Judge Parker was not successful in his attempt to intervene.

As soon as Ned had recovered his health, he began making preparations for defending himself and his wife. They had chosen not to go on the run because the word was out the posse was planning to kill his wife, and she was not in any condition to go on a run. A solid defensive position was his only choice. Ned began construction of a new home. It was more like a fortress than a house. He had double log walls eight to twelve inches thick, with a three-foot gap between the log walls. The space was filled with coarse sand. The house had no windows, only shooting slits. The roof was slate so that it could not be set on fire. The house had a good spring in the middle. It was filled with food storage sufficient to last for many years of constant attack. For more than a decade, he held off posse after posse. The last posse attacked with cannons and dynamite sticks. The concussion broke his wife's eardrums, and thinking she was dying, Ned ran out of his stockade and attacked the posse in hand-to-hand combat, hoping to save her. He was killed, but she lived many years after him and died an old woman.

Ned (and others like him) establish that a home can be built that is defensively sound, though his efforts were likely more than most of us are willing to commit to. Nevertheless, there are many things we can learn from Ned's efforts. First, if your walls are insufficient to hold out against small arms fire, you might as well be standing outside. Ned understood that doors must be heavy enough to hold out against human force. He understood that the windows were totally worthless for anything other than to shoot out of. He also made sure he could return more lethal force than those outside could unleash. With those things in mind, let us look at our own homes and see what needs to be addressed.

DOORS

The doors in your house are critical because they represent the first place any intruder will look to gain entrance. Take a look at the doors allowing entrance to your home. If they are made of wood or of something relatively flimsy, you may need to replace them. My suggestion would be metal doors. Whatever you choose, the material should give you a door that cannot be shattered with something like an ax or sledgehammer.

The next item to be considered is the door hardware—things like hinges or locks. These also will need to be replaced with much heavier materials. Even if an intruder is unable to break through the heavy metal door, he or she can still break through the locks or the hinges. As you look toward replacing the door locks, think about strong deadbolts. Two are better than one. Three would be best. The hinges are of equal importance because they can be taken out with tools such as an ax or sledgehammer or even a twelve-gauge shotgun. You need to take the normal hinges and replace them with heavy-duty hinges. Normally, there are two hinges on the door. You should have three or four to add strength. Now, it should be understood that nothing can stop someone with the right tools and enough time. What you are trying to do is slow them down at the door while you bring in your defensive weapons. In addition to making the intruders work hard to gain entrance, the racket will alert you to their presence. A high-security peephole in the door will also enable you to look through and ascertain who is outside the door. Also, you cannot have doggie doors because they will totally compromise all of your other improvements.

Your front door must be tightly locked and closed. You must be behind that door (unless you are certain of who is on the other side). That means you must see them long before they get onto your front porch. Your front door is the natural chokepoint—the only place where you are in control. If they should ever get through that door, it is all over, and you have lost the fight. Nothing is more important than controlling who comes through the door.

WINDOWS

The next weak point that needs to be addressed is your windows. I understand your love for light. We are all happiest when our houses are full of light. In addition, we feel the need to see outside. Nevertheless, it should be apparent that the glass in our windows is the weakest part of a house. It can be fully compromised by almost anything from a rock to a brick. A BB gun will even do the job, and any larger gun will be even more effective. So how do you get from sight and light to windows that will stop bullets?

You could remove the glass and replace it with steel. However, what about the period of time when there are no enemies at your front door? Now, I'm sure you will want plenty of sunlight, not to mention a nice view. It would seem the best answer to this problem is to have something bulletproof that could be slid in place quickly when the time is right. Something like plate steel shutters or bulletproof glass may be a good option. The shutters should ride on a track so they could be rolled into place. They should have firing slits cut into them so you could fire through them and yet stay relatively safe.

If your house is a split level or two stories, you will need to determine which level you can best defend. In my case, I plan to seal off my basement level and defend the top level because the shooting angle is better. You may find that just the opposite will work for you. My house came with no windows on two of the sides, so I had to build shooting ports on both sides. If you have an attached garage on the side of your house, you will need to be sure that you can see clearly on that side. Your house must be defensible on all sides.

WALLS

Your walls could be the most expensive and difficult to fix if done correctly. No two houses are exactly alike. The problem with doors and windows were enough alike to generalize, but when we take up the problem of the outside walls, we must deal with more obvious differences. For example, if your house is made of large stones, which is to say your outside walls are hard stone several inches thick, you may not have a problem. Bullets will not be going through your walls. If you live in a solid brick home, you're still in pretty good shape. What I'm dealing with here is the typical stick house. In this house, the walls are mostly two-by-fours. The outside walls will be made up of sheets of plywood covered with some kind of siding, probably aluminum. This kind of wall does not do a good job stopping bullets. However, the key to a simple fix is in the space between the upright area that serves as the center of the wall. That space is the height of the wall by four inches deep. There is typically sixteen inches between each two-by-four. The space can be filled with anything from sand, small gravel, broken up pieces of cinderblock, or insulated foam. Many years in law enforcement taught me that one cannot be too prepared when dealing with killers.

To live through these kinds of bad times, you need food. For that food to do what it's intended, it needs to be protected and defended. To be able to properly protect and defend your food supply, you must have it secure so that it cannot be taken away from you. And both you and your family must stay alive so that that food storage can do what it was intended to do. You must control who can come into your home or you cannot protect or defend yourself, your family, or your food supply.

NOTES

1. Federal Bureau of Investigation, "Famous Cases and Criminals: Willie Sutton," http://www.fbi.gov/about-us/history/famous-cases/willie-sutton.
2. Bonnie Speer, "Cherokee Outlaw: Ned Christie," *Wild West* 12, no. 5 (February 2000): 32, http://connection.ebscohost.com/c/articles/2594193/cherokee-outlaw-ned-christie.

10 HOME SECURITY PLAN

"Having a pretty good plan is kind of like having a pretty good parachute."
–Unknown

We are going to assume you already have your food and water stored in a place where you can best defend it. We are going to assume you already have all the weaponry you plan to use to defend it and that you and your family are all trained and ready to carry out the assignments you have been given. We are going to assume your home has been redesigned and made as ready as you can make it for defense. So, what would be next? The answer is a plan for how you will use all of those things in the best possible way. This cannot be something you quickly slapped together. It must be a plan that covers every possible scenario. It must be crafted in a careful, detailed way so that you can sleep well at night, knowing that there will be few surprises. You'll need to know what your response will be for whatever might happen. Why do you think a black ops team can pull off the impossible over and over? It is because each mission is planned down to the smallest detail. What mission will you ever have more important than the survival of yourself and your family? You mustn't leave anything to chance. Because every situation will be different, I cannot do the planning for you, but I can and will point out some key things you will need to give some serious thought to. A good place to start would be in making assignments. You should begin with the general assignments—those things you can clearly and easily foresee.

OBSERVATION POST

Someone needs to be assigned to keep watch 24-7. You will need to decide where the best spot is to keep watch and how long an individual should be on watch at a time. It should not be for more than four hours at a time, especially at night. Either sleep will sneak up on the individual assigned, or he or she will simply become less alert over time. It cannot be stressed enough how important this assignment is, because you must know beforehand when the enemy is coming.

FIRING POSITION

There has to be a shooter on every side of your house. You can almost count on the enemy trying to keep your attention on the front of the house while making an attack on another side. Don't be fooled if at first all the enemies are focused on the front, because the shift to another side of the house may be held off in the hope of distracting you until the time is right for a move to another side.

THE "DO NOT CROSS" LINE

You need to select a line where you feel you can't afford to allow the enemy to come any closer. If there has been any dialogue between you and your enemy, this is the place where that comes to a stop. This will be a line where you, your family, friends, and any others inside with you, understand that the time has come to stop the enemy's advance. You simply cannot allow them to come one step closer. For me, that would be the outside edge of my property line.

WHO IS THE ONE IN COMMAND?

The main reason the military follows a strict chain of command is because nothing else works as well. One person, and one person alone, must be what the enemies fear. A second or third voice clearly indicates disunity and disorder from within. They need to get the message that in this house, they are up against a solid leader and a unified force. The communication from inside must be clear to them so they can make a good decision about their next step. You will be helping their decision if they see your house as a risk too big to handle. You are hoping they will just move on down the block and leave you alone. It is even more important that those inside the house have one clear voice to follow. To have someone else trying to give orders will, at the very least, split the family and weaken them at a time when they can't afford it.

RESERVE SOLDIERS

These are needed to cover the four sides of the house. Since every situation will be different, I cannot tell what you will have left over or how many will be in your reserve. They will need to be ready to move to whatever side of the house is experiencing the most action. Those too young to handle weapons can carry ammunition, communications, or water to those who make up the fighting force.

Let's start with a set of rules we will always follow, especially in the first encounter. The number one rule is *never expose yourself*. For them to do any damage to you, they will need to have a target to aim at. In other words, they can't hit what they can't see. You will need some way to communicate with outsiders without allowing them to see you. However, it is absolutely necessary that you are able to see them, and it is also important that you are able to inflict some damage on them. This must be a situation where communication can happen. The intruders have to know that they're facing serious, determined, well-armed people who are willing to fight rather than give up anything. The first rule in the pillagers' minds is *don't fight unless there is no other way*. They are there in numbers because they're cowards. They want your food because being hungry is not a pleasant feeling. But they don't want to experience a wound to their bodies just to relieve their hunger pangs. They will use every terrorization or intimidation method they can come up with before they take serious action. Nine times out of ten, the problem will end with communication at your front door. So now we will deal with that one time in ten, when the pillagers are not so easily turned away.

The reasons will vary for intruders deciding to stay. The most common reason is that you didn't sound serious enough nor convincing enough. In other words, they determined that you were bluffing. You cannot be satisfied with talking tough if you have nothing behind your threats. Look at this problem seriously. If you have more than one person who could do the talking for you, pick the one most likely to come across as serious. The first encounter is critical, and so much will ride on how seriously you are taken. They will not be able to see you (by design), and their impression will come down to a couple of things: how calm but serious your voice sounds and a preconceived image the pillagers have about you based on what they hear you say. The real problem will develop if they begin to think that you are all talk and no action. If that is the image they have of you, it will be much harder to get them to take you seriously. If you are fighting this kind of image, you may be forced to move to the action mode quicker than you normally would have wanted to. But the longer you leave them wondering, the less likely they will see you as a threat. The minute they decide you are all talk, the standoff will be over, and they will commit to a full-scale attack. That is precisely what you're trying to avoid.

Many will criticize me for saying this, yet it is more or less true that you need to be the first one to draw blood. By doing so, you may save lives on both sides of the attack. You don't need to make that one of your rules in your home defense plan, but it is one of mine. It goes along with one of my other rules: *when the combat is over, I will be the one still standing.* There comes a time in every confrontation when you will know everything has been said. Let me give you a simple metaphor. Two little dogs are barking at one another through a solid fence. They might keep that up all day, but as they run along the fence, telling each other how tough they are, they come to a place in the fence where several slats are missing. That hole in the fence will make it clear which dog is really tough. Half of the time, they will both take off running the other way. Now, in your situation, when you feel you arrived at that hole in the fence, my advice is to start shooting. If that kind of approach is too harsh for you, just say, "I don't believe we have anything left to say, so either start shooting, or turn around and get off my property." Now that approach will leave you a lot more at risk, but you must do what you feel you must do.

If the enemy still wants to talk, one of two things are happening: they are still trying to feel out your strength and build up their courage or they are trying to keep your attention up front while setting up an attack from behind. In either case, continued talk at this point is creating more risk for you and your family. Just imagine you see that hole in the fence coming up fast and bring the discussion to a close.

The real critical part of your home defense plan is the point when the first bullet is fired—whether by you, some member of your party, or by one of the pillagers, it really doesn't matter. What happens next? In this situation, you don't find yourself with many options—only one. You must try to put out as heavy a field of fire as you can. The only thing that will make sense now is convincing them you have them outgunned. If they think they have you outgunned, they will push the advantage until you have given up resistance.

The crime here is breaking and entering. We can also use the term *trespass*, but this might apply to anyone being on your property without your permission. This could apply to several noncriminal situations, such as an old person with dementia, amnesia, or some other medical condition who wandered onto your property with no criminal intent. With breaking and entering, the trespasser has, without your consent, entered your property by force (hence the words *break and enter*). Breaking and entering will not necessarily be a nighttime activity. It might very well be carried on during the day. The lack of law enforcement will make sneaking around at night unnecessary. That will not be a big advantage to you, but at least you will be able to see what is going on, and it will be easier for you to keep your family safe.

Survival will be about how well you can defend yourself and your loved ones rather than hoping law enforcement will help. Your whole life up to now has been about letting the local government take care of you. In the event of a break-in, your plan—in addition to defending your family—should also include the defense of your food supply. If you lose that resource, both you and your family will starve. When you think about it, your plan for survival boils down to three factors: protecting your life, shelter, and food supply. The third factor is of the utmost importance, for if you have no food, the other two won't matter.

You might be alone in your defense. However, if there is a safe and dependable way for you to include others in your plan, then that would make things better for you. In one family I know, the mother and father are older. They have a number of grown, married children. The family has decided to store all their food supplies at the larger, more easily defended home of the older parents. The married children all live close enough to move everything to grandma and grandpa's house and ride out the worst of the famine together. Not a bad plan. Remember, with families at this stage, the mother and father likely put together their food supply when their children were still young and living at home. The food that was stored was a year's supply, and in some cases it was a two-year supply. It was stored not for two but for five or eight people. Now, when the married children bring their food supply and add it to the original supply, the amount may be sufficient. When we think about defending the family and food supply, the additional numbers of defenders will likewise be sufficient. When a group of desperate intruders come pillaging for food or whatever else they can steal, they are looking for a weak person who won't give them much trouble. When they come upon a house defended by several determined fighters making a serious stand, generally the pillagers will move on, looking for an easier target.

Now if this scenario is not possible in your case, another approach might be to include your neighbors. Just be very sure they are people you can trust. Sometimes this might include people you work with. You will generally know them better because you have spent much more time with them. But before you make any invitations, talk to your family. Make sure they are in agreement. It is important that they will all be able to get along well.

What about tactics? First, keep the door closed; speak through a window or with bulletproof shutters in place. If individuals are suspicious-looking, then it is time for the first verbal encounter. It should go something like this: "Go away! This door will not open. I will not be coming through, and you will not be coming in. We have many good shots in here, and they have weapons. If you continue to push your agenda, all you will receive is a hole in your head. You need to move on." This message contains all the information needed at this point. Now, they can take your word and move down the street. There will be no bloody confrontation, which is the best of all outcomes. Or they may not believe you, in which case they will attack, usually by shooting at your house first. The time has come. You need to back up your words. The best shot needs to put a fatal hole in, hopefully, the leader, or whoever is a clear target. When the leader goes down, the rest of the intruders have a second chance to reevaluate their decision to attack. They can move on down the street with only one loss, and you have lost no one.

If enemies come to the side and rear of the house, they are not interested in talking, and they mean to mount an attack right from the get-go. You must have someone watching the side and rear so you won't be caught off-guard. They will come shooting, so you must be ready to return fire immediately. Law enforcement officers prepare themselves for armed confrontation through an unofficial but commonly used motto, which they call the "prime rule." The prime rule simply says that the officer will be going home to his family at the end of the shift. In other words, do not take the kind of chances that will get you killed. This rule is taught by every shift supervisor, and it is the number-one rule.

Whether you are a peace officer or citizen, the principle of the number-one rule remains the same. If the prime protector is killed, where does that leave your family? Your desire to be courageous cannot overrule the knowledge that you must survive the confrontation at all costs. The main question becomes, what can I do to stay alive? The first thing you can do and should do is plan.

To get the most out of prior planning, law enforcement uses something called "scenario training." It involves using mental imagery to create similar situations to real-life events. If the situation is pictured vividly enough in an officer's mind, his or her brain is able to use that imagery as if the situation was real. With traumatic experiences, the brain begins to sort through your mental memory file for some experience to help it know what to do. If there is nothing there in the files, the brain goes into a kind of a mental lockdown, kind of like your computer when it fails to find what it needs to move ahead. Few people have had experience anything like a TEOTWAWKI scenario. But when officers' brains know what to do through mental imagery, they simply move into that "previous experience" and bring the situation under control.

As a peace officer, I'd drive along, covering my assigned routine. As I did, I would put my imagination to work. For example, I would try to imagine a vehicle stop. I'd pick some car out in front of me and imagine seeing a major violation, making it necessary for me to pull the car over. In my mind, I would see them pull off to the side of the road. I would then carefully go through each action in our policy: I would park my car at an angle so I could use it to cover me; I would call the dispatcher and give him or her proper location information and the license plate number, color, make of the vehicle, and how many people were in it. When I had fulfilled all the protocol listed under departmental policy, I would exit my vehicle, clipboard in my hand, and service weapon unhooked and ready. As I made my approach, I would stay tight against my vehicle, and when I reached the pulled-over vehicle, I would make sure the trunk was shut tight and locked. Looking through the windows, I would make sure I could account for everyone I had originally counted in the car. If anyone tried to exit the car, I would order them back inside. When I began my conversation with the driver, I would keep my clipboard between us while watching his hands, never allowing them to get out of my sight. With any trips I needed to take back to my vehicle for radio confirmation, I would follow the same routine.

The problem is an officer can do this one hundred times a day and everything will go just fine. Then, after the hundredth or two hundredth time, he or she becomes lazy and careless. Why do I pick this situation to role-play in my mind? Because more officers get shot while making vehicle stops than in any other routine activity. Mental scenario training will also be helpful to you as you prepare to defend your home.

Let's walk through your situation again. First, your house is your castle. The house is also the battleground where you will fight. The front door is the chokepoint. It is the tipping point as to whether or not war will be declared. It is the one place where you will have some influence on what you will or will not be able to do. Take time to go over all the possibilities, and you will begin to see options you didn't realize you had. By identifying problems, you will also fix them before they become actual, real-world problems.

The next thing you should do is stand around that key area and start working out as many different scenarios involving different forced entries as you can come up with. What would your enemy do? And what would you do to counter it? The beneficial thing here is to make sure nothing is going to happen that your brain hasn't seen before. The last thing you want to happen is a mental lockup on your porch. As long as your brain feels it knows what's going on, it will continue to function in creative ways. Once you feel you have exhausted the many possible scenarios that involve your front porch, the next thing you should do is to repeat the same process throughout your house. Where are other places an intruder might try to gain access?

It's your house, and you have all the advantages. You know the battlefield. You should be able to anticipate an intruder's every move and be able to design counterattacks that they will not be able to anticipate. In the early time—that period of time when people are just beginning to realize that all the food in the market is gone and may not be coming back—the first pillagers will most likely be small groups, generally two or three. It would be rare for one man to be gutsy enough to come alone.

Once you have made a strong commitment to yourself, you're ready to begin your training. The first thing you need to focus on is a comprehensive, all-inclusive plan. This we will call your "home security plan." It will be helpful to get the whole family involved designing the plan because a member might see something the others missed. There are several parameters you will need to focus on. For example, everything in the plan will be focused on your home. That will be the castle, the place you will be defending. That is the battleground you will be working within.

If you carry your defense outside your home, you may have made a fatal mistake. Inside the home, you have the advantage, but allow yourself to leave that protection, and the enemy has all the advantage. Rule number one—*never leave your defensive position*. Do not even step onto your front porch. Next, place your food storage somewhere in the center of your defenses. Lose that, and the rest is moot. The house must be able to be defended. It's your castle, and it must be able to repel, hold off, or repulse any enemies trying to force their way in. You must hit back with equal force so you can you force them back.

You cannot be casual about protecting your food supply; otherwise, you may as well not store any food at all. Further, a casually prepared home defense plan is as bad is no plan at all. The idea behind a good home defense security plan is to head off all the ways people with bad intentions could catch you off-guard. You need to plan ways to keep unwanted people outside. If you don't keep them on the outside, you may lose, no matter how well you prepare.

If you have chosen to bug out, you still need to be ready to defend your dwelling. You might be planning to live out in your bug-out vehicle or in some kind of cabin or other dwelling you have constructed. Either way, everything I've said about defending your home remains the same. You are defending your new home. This will be like Ned Christie's defense. There will be a few things that will change, such as the distance in the firing lanes. For the bug-in folks, shooting distances are short, usually no more than fifty yards. But with the bug-out folks, it may be as far as you can see. Consequently, the choice of weapons will need to change. Then there is that line you don't want the enemy to cross. It will be further out. In fact, you will not want people to come any closer than it will take to identify them.

If you are in the process of building a defensible home, you will not be thinking in terms of a present-day model. You can build it right from the beginning. Like Ned, build it with defense in mind. Build it with all the advantages going to you. Build it around your personal security plan. Build it heavy enough to withstand a serious attack.

11 HOME DEFENSE WEAPONRY

> "The best argument for peace . . . is nobody wants to shoot if somebody is going to shoot back."[1]
>
> –Louis L'Amour

When a large number of individuals are standing in your front yard, armed to the teeth, with the clear intention of killing you and your family, or at least taking all your food and water, it would seem to be a just cause for pulling out a weapon and, if necessary, killing those who attack.

As a retired peace officer, I am particularly sensitive to the attitude people have concerning guns. It is clearly a subject where people have strong opinions. My curiosity on this subject led me to question many people about their reasons for their positions. My unofficial surveys did not produce definitive answers, but I have noticed (even though admittedly this is a weak generalization) that there seems to be a notable difference between age groups. It seems older folks, those fifty and up, are those more in favor of guns, whereas the younger group are more likely to be anti-gun. It's those on the low end of this group, those in their twenties and late teens, who are aghast at the idea of owning a gun. If someone were to ask me how to explain that disparity, my best guess would be the media. The anti-gun movement has always had the television as its biggest voice. In addition, those under forty are less likely to have ever owned a gun or even fired one. Many people of the generation over forty grew up hunting, which meant that they had an opinion based on personal experience. Those with personal experience with guns are not going to fear them and are going to have better feelings about them. The younger group generally seems to see guns as evil things. To the older group, guns are just tools. There is an exception to these trends, however, that runs through the age-group theory, and it is those who have served in the military. That includes men and women. For them, a gun is a tool, and usually in their minds, the one thing that kept them alive.

Whatever your views on gun ownership, when considering protecting and defending your home and your food, there isn't anything else that can work better than a gun. Massad Ayoob, one of the masters of self-defense, noted that the role of the firearm is protection: "The firearm is the only really effective means of defense against a vicious homicidal assault. Nothing else can serve this function of the individual law-abiding citizen."[2] When desperate people come to your house with the intention of taking your property, be it food, clothing, or something even less moral, you can be sure they will come armed. Clearly, their purpose will be to overpower you, and you can be sure they will come armed with lethal weapons. If you have hope of protecting and defending your family, you will need to be ready to fight fire with fire. On the street, the same notion might be expressed as "only a fool brings a knife to a gunfight." The only way you could hold your own against an armed assailant is to be armed yourself.

A private citizen has much in common with peace officers the moment the citizen makes the decision to arm himself or herself. Just a week ago, a good friend of mine informed me he had purchased a handgun because after talking to me he decided it would be a good idea. I didn't feel like breaking the bad news to him, but so far, all he had purchased was a false sense of security. Buying a gun hasn't changed anything for him other than he now thinks he has a new security. Yes, now he has a gun, but he doesn't have any idea how to use it. Can he get at it when he needs to? Does he know how to load it, find the safety, and fire it? Could he hit anything with it? Buying a gun is similar to buying a Stradivarius. With no musical training, the best you could do is look at it. And while it's a beautiful instrument, it is meant to be played to give the world beautiful music. Just owning it would be a waste. In a similar way, a gun is useless sitting in its original box, still wrapped in paper. It is intended to be used. After you have taken it out of the box, learn how to properly maintain it, load it, and shoot it. Shoot it well and accurately, and then you can begin to take pride in the ownership. Then comes additional hours of learning how to conduct yourself in a life-and-death situation. Buying a gun is a little like paying the cost of the class; after paying for the class, now it's time to start learning the subject material.

An account given in Chris Bird's book *The Concealed Handgun Manual* illustrates the value of gun ownership and, at the same time, much of the country's attitude against guns. At the time of this story, Chris was the sergeant on duty. A call had come in from the parents of a ten-year-old girl indicating that a rape was in progress. The parents had been visiting some friends down the street, and when they returned home, they went to their daughter's room to check on her and found her being raped by a young man, who as it turned out, lived only four houses away. The rapist managed to escape out the bedroom window. The parents had the right mind and immediately called 911. As it turned out, this was one of those rare times when police were close by. The police had lights on the rapist before he could get to his own house. However, instead of giving up, he pulled out a 1911 colt .45 and began to shoot at the two officers. One officer managed to get a 9 mm round into the rapist just above his belt. The officers then called an ambulance and transported him to the nearest hospital.

Sergeant Chris showed up at the hospital to check on his men and then tried to engage the hospital staff to ascertain the condition of the rapist. They acted as if they were angry with him and the other officers. There were two nurses and a young doctor. They were upset because the nice young man they had been working on was a victim of "police brutality." He clearly was in great pain—belly wounds tend to be painful—and they got caught up in his pain and in his insistence that he didn't do anything wrong. And believe me, the worst criminals see themselves as the victim. The medical team had completely forgotten that this nice young man had just been caught in the act of raping a ten-year-old girl and that it was this "nice young man," not the cops, who it initiated the gunfight. The cops were returning fire in defense of their own lives. When Chris reminded them of those two important facts, all they could think to say was, "Guns! Who needs them?"

Chris responded, "The good guys. The bad guys already have them."[3]

Using a gun for home defense is the only logical way to make a successful stand. As noted in Chris Bird's book, many people who normally wouldn't own a gun, when confronted with realities of self-defense, often change their views to pro-gun. Bird cites Eliza Sonnyland, a talk show host for KTSA Radio, San Antonio, Texas. She is not a native Texan. She grew up in New York, living in Manhattan on Long Island. There, she had no exposure to guns. "There were no guns. There were no conversations about guns," she said. At one point, she lived in a particularly unsafe area of New York City. She carried mace, but it never occurred to her to protect herself with a gun. As an adult, she absorbed information from the media about guns and it was all negative. "What am I getting from the news and information out there? Bad guys are shooting people. Kids are accidentally getting shot. Famous people are being assassinated. What is the message? We got to get rid of guns. It's a given."[4]

She then moved to Texas, where there are lots of guns. She arrived in San Antonio in June 1983 to visit a college roommate and never went back. She got a job as a radio producer, became the host of a morning talk show, and later got her own radio show. The first time she actually held a handgun was when the public information officer with the San Antonio Police Department insisted she go to the police range and shoot. "I never felt the weight of a gun in my life," she said. She was impressed by the way Sergeant Paul repeatedly checked the handgun to ensure that it was unloaded, even though he knew it was unloaded. He showed her how to dry fire the gun before she was allowed to actually shoot it on the range. When she hit the target, she felt an unexpected thrill. Sonnyland said she grew up with a negative view of hunting and hunters. She was even a vegetarian for part of her life. "I still believed that hunters were nothing but a bunch of sadist killers. They got their jollies from killing things. I mean, who are these people? And why would anyone want to talk to them?"

At the time, she was living on three acres outside the city, and she kept chickens. She was upset because the raccoons were killing off her chickens, biting off their heads. "They seem to kill them for the thrill, and all of a sudden, I don't want to drive to work, swerving around raccoons. I wanted to run over them. All of a sudden, I have this different look on nature, something I never would imagine would happen." One talk show guest provided her with more food for thought. He said that there is no quiet, gentle death in nature. It tends to be brutal and terrifying. "I had never thought about that," Sonnyland said. She realized that short of deliberate torture, anything humans do to animals while hunting is more likely to be quicker and less terrifying than the way nature disposes of the old and infirm. That was a big shift in viewpoint, she said.

However, a bigger shift was yet to come. "I was having one of my 'I know more than anybody else' kind of conversation with listeners about guns and why we must do something, and a woman called me up. She was an elderly woman, and she said, in a very matter-of-fact voice, without being emotional or scolding me, that she had been raped, and she now carried a gun. That was it. Who was I to tell her what she should do? Well, if I'm going to accept that nobody can tell her she could, who was I to tell anyone else they shouldn't?" The next attitude change happened when Ralph, the owner of a local shooting range, persuaded Sonnyland to take the Texas concealed handgun licensing course. After missing several classes and being prompted by her boss, she had a change of heart and finally turned up to take the course. She was surprised to find that the others in class were normal people with lives and jobs. She was impressed with what she was learning. "I'm learning through this concealed handgun course the last thing I want to do is pull a weapon out and shoot somebody with it. I'm thinking, why isn't everybody taking this course whether they intend to use it ever?"

The result of all these experiences and her willingness to open her mind and question everything she had believed was when her attitude about guns changed completely. She took a three-day defensive tactics course at Thunder Ranch Shooting School and took up cowboy-action shooting. She now goes by the name Sunshine Kid. Now when she interviews people about guns, she tries to remember that it took her years and many experiences to change her mind. Sonnyland urges shooters to invite others to go shooting and let them feel the thrill of hitting the target, as she had. "It's the hands-on experience that will give someone an education that no amount of argument, verbal abuse, or anything else will do."

I served as a range officer for nearly twenty years. This role was not just about directing peace officers and teaching them what they needed to improve to qualify on the range. In some cases, the role included helping officers gain a positive feeling and a new respect for firearms. Most peace officers have been hired as officers of the law, so it was understood and expected that they would be experienced with firearms. The challenge here, though, is to get someone who has no such requirement to agree to learn how to use one and to do it with a decent level of skill. Many times, I ran into individuals who would say something like this: "I will agree to keep a gun in my house and to use it in my home, but I don't want one that will kill someone. I just want to wound them enough to stop them and get them to leave my home." First of all, few people really want to kill another human being. But there will be times when the enemy gives no other option. It is important that you understand that wounding someone will rarely stop them. Wounding seems to be a Hollywood thing, but I can assure you that the things you have learned in movie gunfights are cinema mythology. They will not work in the real world. Let's start over and see what we can learn together.

We can forget about the aggressive side other than to try to anticipate our opponent's movements. The issue is aggression versus defense. In the survival scenario we have been speaking of, pillagers are the aggressors. They will be the ones bringing the fight to you. You will almost always be the defender. You will always be fighting from your home, your castle. You will have no need to learn the tactics of the aggressor other than to understand the battle from his or her point of view. What you need to understand are the tactics of the defender.

The next issue after acquiring a gun is knowing if you intend to use it. A while back, a friend of mine told me, "As soon as I get my concealed firearm permit, I'm going to get a handgun so I can defend my home." Clearly, he was confused about the legal requirements. As long as you're using your handgun to defend your own home from intruders who are intent on committing violence or a felony, you are legal according to something called Castle Doctrine.[5] In your home (your castle), you are seen, in effect, as a soldier with all the military rights to do what at you have to do to defend your castle. Another difference is that the person coming into your castle has crossed a line—by legal definition, intruders are trespassers who waived any rights to live when they broke into your home and threatened your life.

For many readers, guns are new. Other readers, however, are probably well versed and have an advanced understanding of guns. The following information is provided as a general primer and is designed to help those who are less familiar with firearms.

Should you choose an automatic or a revolver? Let's take a quick look at the features that may represent an issue in each. A minus for the revolver is the amount of shots available with each track or reload. Most revolvers have a cylinder designed to hold six rounds, so you can shoot six times, and then you will need to reload. Unless you are interested in playing cowboy, most revolvers today are double action, which means even though there is a hammer on the outside (like the old single action), you must pull the trigger to get the round to go off. One of the biggest advantages of the revolver, however, is its simplicity. A revolver is not designed with a safety; It is "hot" out of the holster. Whether that's a good thing depends on your level of confidence. An automatic, in contrast, has a larger capacity (usually twelve to fifteen rounds, depending on the caliber) and is more complicated to operate. Automatics often have a safety, and all automatics have a slide, which must be pulled backward—toward the shooter—to load the first round.

People who like revolvers tend to swear by them. I used to be one of those. I still love revolvers, and I own a bunch of them, but I have learned to love automatics (at least some automatics) just as well. I am personally not a big fan of safeties, so even my automatic must be hot out of the holster. My wife, on the other hand, is a lover of the revolver. She does not own an automatic. When deciding whether to buy a revolver or an automatic, don't buy in a hurry. Most cities have shooting ranges where you can rent time on their range. The range masters will bring you a selection of revolvers and automatics to try. Make sure you like the gun and that it feels good to you. Too many people have bought a gun because someone else told them the gun was better than any others. If you choose to listen to another's advice too closely, you may not end up liking the gun you buy. Remember, everyone is an expert, at least in their own mind, so you end up with their preferences rather than real performance. You should understand that, like almost everything else, better quality costs more.

What caliber should you buy? All calibers have a name or nomenclature, which, for the novice, can be confusing. The US designs its calibers differently than Europe does. In the US, a caliber is established by the distance between the barrel lands, or the helical grooves, also known as rifling, on the inside of the gun's barrel that gives a bullet its spin. This distance is measured in inches, hence the terms .32, .38, .40, .45, and so on. In Europe, the distance is measured in millimeters, which gives you calibers such as 9mm and 10mm. It gets more confusing when other terms are added, such as *Magnum*—as in .44 Magnum. There are other confusing terms such as the .45 Automatic Colt Pistol (ACP), which is a smaller .45 used in an automatic. Sometimes it helps to look at the size and shape of the bullet, but asking an expert or doing some serious reading may be more useful. The best thing is to do some hands-on research. Shoot a variety of today's more popular cartridges and see for yourself. The chief issue here is getting the biggest caliber you can comfortably handle.

The smallest caliber I recommend is a .22, and as strange as it may seem, this caliber is popular with professional assassins for its lethality and low noise. In my opinion, however, the correct approach is to purchase a handgun with a cartridge large enough to knock down the perpetrator. When intruders are knocked off their feet, their momentum is stopped, and when their momentum is stopped, it will be harder to get back into attack mode. This will give you time. Not much, but just a little to do what the situation will call for next. Most peace officers (including me) would advise you to keep shooting until the intruders can't get up, but the decision is up to you. The human body can take a surprising amount of punishment and still constitute a threat to others. In the movies, people drop, roll over, and die when they get shot, but it doesn't work that way in the real world. I was a boxer in middle school, high school, and in the army. My coaches taught me, "When you get him down, don't let him back up." This is doubly true in a gunfight.

In present-day law enforcement, the general rule is to have nothing smaller than a .38 caliber gun. Occasionally, a detective will opt for a smaller caliber only because he or she doesn't plan on having to use it. Smaller guns also mean lighter and more easily concealed guns. However, anyone who thinks he or she may get caught in a shootout should opt for a heavier caliber—both in weight and, most important, in knockdown power. Too often, peace officers are saddled with a choice made by a suited-up line commander who sits behind a desk. More often than not, the commander has not been in a firefight and has no idea what that entails. Let every officer have his or her own choice, and the calibers would go up. What, then, is the best weapon for home defense? First, to be adequately armed, you will need more than one gun. If you are forced to buy just one gun, my advice would be a good handgun. Remember, you are concerned with a defensive stance only, and when you are shooting from within your house, the range or distance will be short. Generally, most handguns will perform well up to fifty yards.

Caliber is about the type and size of the bullet your handgun shoots. This is an area where many people make dangerous mistakes. Again, go to the shooting range and try a variety of cartridges. Test each caliber's recoil, also sometimes called a kick, and see if you can control it. Both the gun and the cartridge contribute to recoil. If the recoil is too severe, then you will never learn to shoot accurately. It is more important that you hit what you are shooting than brag about how big your handgun is.

While working as a peace officer, our duty piece was a 9mm Glock. Then, for reasons originally unknown to me, the department's director thought it would be good to trade our 9mm's for .40 caliber Glocks. I loved the change. The .40 hit a lot harder and was more likely to put the perpetrator down, though the recoil was more intense. After the change, though, officers who had done well with a 9mm, sometimes shooting among the highest scorers, were finding it difficult to shoot well enough to pass grade. I petitioned the director to allow these folks to go back to carrying the 9mm. The director was hesitant to allow exceptions to the .40 caliber change because, in his opinion, if our officers carried the same cartridge, they could swap ammunition in a shootout. If an officer ran out of ammunition, he or she could call to a partner and have an extra clip passed over. This was a sound reason for ammunition standardization. However, some of our officer's weren't strong enough to handle the .40 caliber, so some officers were allowed to carry a 9mm, and the rest carried the .40 caliber. Similarly, you will need to sample the various calibers to determine what level of recoil you can handle comfortably.

If you are the only one operating in the defender role, one good high-quality handgun should be enough. However, it is best to try and cover every contingency, and owning several guns might be better. Even high-quality guns wear out, break down, jam, or lock up and require repair. In addition, what do the other members of your family do while you are making a defensive stand? You will do whatever seems best to you. You know the other members of your family better than anyone else, but one of the hardest things to do is huddle in fear in a corner of the house while someone else is taking on the danger. Generally, it is better to be busy helping deal with the danger. In my family, I purchased a gun for each family member when they came of age. I took them out to the shooting range and taught them everything needed for handling that gun safely and effectively. They were trained to know that they would help in the defense of the family. One defender can realistically defend only one side of the house at a time, giving pillagers better odds of making it in. If there is defensive fire coming from every side of the house, though, the odds are better that no one will take a bullet.

We have looked at the need for a good handgun, and I have suggested my preference for a backup handgun in case the number-one handgun breaks down. I have left the choice of the revolver or automatic up to you. We have considered the cartridge, also known as the bullet. I have suggested that, unless you are an assassin, you might do well to stay away from little bullets like the .22. It is worth noting, however, that a .22 is hard to beat when trying to teach a child how to shoot a handgun. A .22 loaded with stinger hollow points can also do a passable job, especially for someone having difficulty with handgun recoil. A solid hit with a .22 beats a clean miss with anything else. When the shooter shows the ability to adjust to recoil issues, my suggestion is nothing smaller than a .32. When it comes to caliber, I generally recommend going as big as you can genuinely handle. For a tight group at twenty-five yards, you should be able to cover the group with a dessert saucer. It is important that you be able to hit what you are shooting at.

The gun's holster is also important. If you are forced to maintain a defensive posture for several days in a row, you will wear your handgun 24-7. Pick a gun and a holster that will be at least somewhat comfortable for long-wear situations.

I know there is a lot more you would like me to cover about the handgun, since it is the backbone of your home defense system, but unfortunately we cannot always control the battle for your castle. While it is generally true that the distances between the combatants will usually be short, that cannot be guaranteed. If the terrain around your house makes combat from longer distances a possibility, you will need to find some kind of long-range gun.

THE LONG GUN RIFLE

All of my married children have chosen to address the long-gun issue in different ways. Because we all live in a part of the country that encourages big-game hunting, most of them have opted to solve the problem of big-game hunting and home defense with the same tool: the standard deer rifle. Several of my family members have also purchased military assault weapons, such as civilian versions of the military's M16, which provide more firepower.

THE SHOTGUN

Another piece to consider is the shotgun. There are so many firearms designed for home defense, it is mind-blowing. Most experts agree the tactical shotgun is the best choice. It's shorter and easier to handle in confined spaces like your home. It's easier for a novice to use; it's a point-and-shoot gun. In trying to defend your home, I couldn't recommend a better weapon.

It is generally thought to be a short-range weapon more in the range of the handgun, but it puts out a lot of lead at once, and as the distance increases, the pattern or spread of the shot widens. This means that within a given range, say ten to fifty yards, you can hit what you are shooting at without careful aiming. So when you need a weapon that calls for pointing and shooting, the shotgun is the weapon for the job. Ideally, you should look at a gun between a sixteen to an eight gauge. My choice is a twelve gauge, which is plenty powerful. Because it is the most popular shotgun gauge, ammunition is readily available. Because you will likely be using it inside the house, the shorter the shotgun, the better. The size of the shot or pellets will be a consideration as well. The smaller the shot, the more each shell will contain, but as the target gets further away, the smaller shot loses effectiveness. Most law enforcement officers favor the two-shot, buckshot, or double buckshot. I prefer BB-size shot as a good compromise.

MILITARY LONG GUNS

We often study history to keep us from repeating the mistakes of the past. Why does anybody need an assault rifle with a high-capacity clip? The anti-gun followers continually state that no one needs such a rifle. Is that the only thing our founding fathers had in mind when they made the Second Amendment? The single biggest reason for it was home defense. The people of this new country were surrounded with dangers.

Without firearms to defend themselves, the colonists would have simply been rolled over by the English, and the movement for liberty would've been over before it got started. Why do so many people in today's world fail to see our need is little different? We are under attack from everything from break-and-entry criminals to rapists. As we look ahead to an end-of-the-world scenario, there is little question that an assault rifle is the perfect weapon to deal with looters bent on ripping your house apart. In the last decade, millions of these weapons have been sold in this country to folks who had home defense on their mind—which is precisely what our founding fathers wanted. Do you need an assault rifle with a high-capacity clip? Yes. I only hope that by the time you are ready to make such a purchase, you will still be able to find such a weapon on the market. The movement trying to take them out of your hands is an extremely powerful lobby. They have succeeded in doing just that in Europe. They have almost succeeded in New England.

EDGED WEAPONS

I suggest that you stockpile all the ammunition you can acquire. If you think about it, you will realize that when you run out of ammunition, your thousand-dollar rifle is nothing but a heavy stick. Buy as much as you can afford for each weapon. I understand how expensive it is, but the price is going up almost daily, and who knows if ammunition will be available in the future? You should also think about buying loading equipment to keep you supplied longer. However, when you run out of ammunition, what will you do next? Having a few knives, swords, or other edged weapons around is a good backup plan.

I have a good prepper friend who has some serious ideas about edged weaponry. Though he has many firearms and stockpiles of ammunition, he sees the need for edged weapons. For example, he has a separate machete stored at each window in his house because he wants a weapon on hand if somebody with bad intentions sticks his arm through the window. The machetes are relatively inexpensive—a good, high-quality South American machete costs as little as twenty dollars. The machetes serve a dual purpose too. They also function as a tool for clearing brush and similar cutting tasks. Machetes are relatively light with a thin blade and are heat treated to bend, flex, and handle abuse. They also take a reasonably sharp edge and are easily resharpened with a file.

My post-retirement years included a part-time job selling knives. It never ceased to amaze me how little the average customer knew about knives; most seemed interested in only what the knife looked like. But the real issue is the quality of the blade—what steel is the blade made from, what is the blade's edge profile, and how well is it heat treated? These factors, when combined, determine how sharp a blade will be, how long it will hold an edge (or how long it will stay sharp during use), and how easy it is to resharpen.

There are numerous kinds of metals used to make knives. Knife steels vary in the materials they include, and the various steel recipes are not unlike the recipe for a cake. Different ingredients must be used in proper amounts, added in the proper order, and then baked at the proper temperature to harden the edge. Modern blade steels tend toward stainless steels, which include ingredients like chromium to make the steel harder and more resistant to rust or corrosion. Blade steels are numerous and include cheap, no-name steel to stainless steels like 420, 440, 440C, D2, and dozens of others. These types of steels vary, and many, such as 420 and 440, are generally considered low-end steels. There are also steels called stainless that are really more stain *resistant* than stainless. These stain-resistant steels tend to take a better edge and retain it longer while staying relatively free of corrosion or rust. Many stainless steels, though, have a reputation for being hard to sharpen. Stain-resistant steels include AUS-6, AUS-8, CPM 3V, CPM S30V, ATS-34, and several others currently popular in today's knife market. Several of these steels, when properly heat treated and ground, make high-quality knives.

Numerous knife manufacturers use high-carbon steel (1085, 1095, and so on) to make knives much like knives and swords have been made over the last several thousand years. High-carbon knives lack ingredients (like chromium) to make them stain resistant or stainless, and as a consequence, will rust if not properly cared for. However, knives made from high-carbon steel, when properly heat-treated, will carry an edge and are relatively easy to sharpen. Damascus, a very old steel type that dates back 500 years before Christ, is made by combining several types of high-carbon steel and is currently seeing a surge in popularity. US-made Damascus is generally high-quality, though Damascus is also made in high quantities in countries like Pakistan and India, and the quality of this Damascus varies from junk to acceptable. The prices and quality associated with these steels vary, so do your research when looking to purchase stainless steels, and pick steel that meets your needs. My preference for a multiuse knife is a good high-carbon steel or a good stainless tool steel. High-carbon steel will serve well in multiple functions, and it's not overly expensive.

Small knives have their place in the prepper's toolkit. The smallest knife is what is commonly referred to as a pocketknife, or a folding knife. The folding knives of today are a big improvement on the earlier stockman's pocketknife (though these classics are still manufactured and carried by many knife enthusiasts). Many folding knives have a clip, allowing you to attach the knife to your pants pockets, where the knife is readily available and secure. Modern folding knives also have locking systems to keep the blade from closing on your fingers during use, as well as a thumb stud or flipper mechanism on the blade to allow quick deployment. Most folding knives have a blade around three inches and not more than four inches. Though this size might sound inadequate for self-defense purposes, law enforcement statistics show that more than the majority of knife fights are fought with knives in this size range. Though not an ideal self-defense weapon, a folding knife can function well in that role when needed. However, in a world with no electricity or gas, folding knives will more often function in a utilitarian role. Utilitarian tasks are daily, innumerable, and include whittling wood into various tools, processing animals, gathering wild plant foods, or making snares.

Similarly, longer fixed blades will prove invaluable during TEOTWAWKI—both as utilitarian tools and self-defense weapons. Generally, carrying multiple knives for each task could be cumbersome, so a fixed-blade knife should be designed as a general-purpose tool and be able to function equally well as a working knife or a fighting knife. The blade should be five to twelve inches long, with shorter knives more able to handle fine cutting tasks and larger knives—generally more than seven inches—more capable as chopping tools.

In the closed conditions of the average house, the knife you have on your person will likely be your go-to knife when fighting is called for. This means the blade should be designed to serve as an effective fighting knife. Thanks to the many years of combat this nation has been involved in, there is an abundance of knives on the market for just that purpose. Most mixed-use knives (utilitarian and fighting) are single-edged. Using a single edge blade allows for more generous angle and a sharper knife, as well as allowing the back of the knife to be used with Ferro cerium rod, magnesium, or flint to generate sparks for starting a fire. You can also hammer this type of knife through wood (by hitting the back edge of the blade with wood) to split it into firewood, though an axe is a better tool for this kind of task, and hammering a knife is generally not recommended. Many multi-use knives have a false edge ground near the tip. This is to increase the blade's penetration abilities. Some false edges are sharpened, but often they aren't. Combining a single-edge knife with a false-edge is a good compromise and provides a knife capable as a tool and a weapon.

Generally, a knife that is full-tang (or a blade and handle that is one solid piece with two handle slabs pinned or bolted to the handle) is generally considered the strongest knife type and the least likely to break. Half or stick tangs are acceptable if well made. The knife handle should be ergonomic and comfortable to use over an extended period. It's also helpful to use a handle material that provides traction, or grip, when the handle is wet. Common handle materials include wood, stabilized wood (resin-impregnated wood), or man-made materials like Micarta, G10, or rubber. A handle that is shaped to allow you to easily index the blade, or identify the orientation of the point and edge without looking at it, is also useful.

When choosing a knife, don't get distracted by terms such as *survival knife*, *combat knife*, or *tactical knife*. These are just marketing terms. Put your hands on a variety of knives and pick one that feels good to you.

A sword could also be a good addition to the home-defense arsenal. Swords are longer than knives and are harder to use in confined spaces. Swords, like knives, come in many types and shapes and range from shiny, decorative wall-hangers to grungy, combat-ready blades. The most important factor to keep in mind when picking a sword is quality. If you're planning to purchase a sword, you'll need to be sure it has good steel in the blade. Try to deal with companies who guarantee their swords will be battle ready and avoid decorative wall hangers. Look to fully functional companies like Ritter, Cold Steel, and Knights Edge.

BOWS AND CROSSBOWS

Today's bows (both traditional bows and crossbows) are more advanced than the primitive bows of our ancestors. Modern bows are more powerful, have greater range, and are more accurate. With the addition of scope sites to a crossbow, the crossbow nears the lethality of a rifle, without the noise. While they are a little unwieldy in the close confines of a house, they are nevertheless important when you don't want to draw attention to yourself. Bows and crossbows also use recyclable ammunition—arrows and crossbow bolts can often be retrieved and used more than once, which is a useful trait in TEOTWAWKI scenarios.

AXES AND TOMAHAWKS

Axes and tomahawks are some of man's oldest tools and weapons and make worthwhile additions to the prepper's arsenal. Axes and tomahawks are readily available and, in most cases, are relatively inexpensive. Like knives, axes serve a variety of purposes and function as a utilitarian tool and, when needed, an effective weapon. Axes and tomahawks come in many shapes and sizes, ranging from large wood-splitting axes to small, nimble tomahawks. Current trends in the cutlery scene include full-tang, one-piece metal axes, and tomahawks designed for breeching doors, chopping holes in walls, and generally smashing and crushing without breaking. Wooden handles, once broken, can be replaced by buying replacement handles or carving them, though all-metal models more easily avoid the pitfall of breakage and wear-and-tear. Examine a variety of axes and tomahawks and pick one or more to augment your kit.

STONE TOOLS

When all else fails, if we find ourselves with nothing but the clothes on our backs and without metal tools, one simple option remains: stone tools. When needed, a stone knife can be manufactured from flint or obsidian. If these materials aren't available, the bottom of a glass bottle works as a survival tool. When shaped, these glassy materials can be made into sharp cutting tools. With a little practice, you can learn how to make usable, sharp flakes for processing game and a host of other cutting tasks. Stone tools are also usable as a last-ditch weapon.

NOTES

1. Louis L'Amour, *Callaghen* (New York, Bantam Books, 1972).
2. Massad F Ayoob, *In the Gravest Extreme: The Role of the Firearm in Personal Protection* (Concord, NH: Police Bookshelf, 1980).
3. Chris Bird, *The Concealed Handgun Manual* (San Antonio: Privateer Publications, 2011).
4. Ibid.

5. 213th Legislature, State of New Jersey, Assembly No. 159, May 6, 2008, http://www.njleg.state.nj.us/2008/Bills/A0500/159_I1.PDF.

12 FRESH PLANT FOOD

> "When pioneers, prospectors, and others later began daring the plains and deserts, many of them starved amidst abundance because they didn't know what to eat or how to prepare it."[1]
>
> –Bradford Angier

I recall seeing a full-page photograph in an old *LIFE* magazine showing a large number of starved, war-ravaged survivors sitting in a field, waiting for the Red Cross to bring them food. The survivors were severely starved and many were near death. I have been teaching survival skills most of my life, and so the photograph saddened me more than it would a non-survivalist. The starved people were sitting in a huge field of pigweed, sometimes called lambsquarters, which is a member of the Chenopodium family. It makes a wonderful salad, is a great pot herb, and is very palatable raw. These people were literally sitting in a field of food but were starving to death. Generally, there is an abundance of plant food growing all around us, like pigweed, which we neither see nor are aware of.

Just because we can no longer go to the supermarket to buy a head of lettuce doesn't mean we can't put together a fresh salad. A salad made from readily available plant foods—many of which are considered weeds—will help stretch your supplies and should be included when available. Finding fresh food to augment what you have stored greatly increases your odds of surviving a famine. One of the problems in living off food storage is that, because of the need for a long shelf life, food storage is dry. You know what it consists of: dry wheat, dry corn, dry beans, sugar, powdered milk, and so on. Finding fresh food on a regular basis, in addition to augmenting your supplies, will also break up the boredom of dried foods. We do not know how long TEOTWAWKI will last, but it is wise to know the plants that might be growing abundantly in our own yards. The more you know about wild edible plants, the greater your chances of survival. We will need it all—a well-stocked supply of dry food, familiarity of wild edible plants growing around our homes, and a wide-reaching knowledge of wild edible "living" protein we can harvest.

We have turned our dwelling into a fortress. We have food and water. We are armed, trained, and ready. After you have been living on your stored food for a while, you will find most of it to be dry. You might be starting to miss a fresh salad. But don't despair, there is a remedy at hand. If you have chosen to be a bug-outer, you are probably in a more remote area, so your plant food choices will be better. People who bug in, however, still have options and should be able to find enough fresh greens to stay happy. The plants growing in your own yard, especially, can add fresh, nutritious food with much-needed fiber. Your unkempt lawn will be especially helpful. Because you probably have not mowed it during TEOTWAWKI, it can provide substantial amounts of salad. Who would have thought someone could be rewarded for being a less-than-perfect gardener? Here's a taste of what you can harvest from your own yard.

GRASS

I mean your lawn itself—that grass is edible. Because you won't likely be mowing it during TEOTWAWKI, it will likely be long and abundant. It should be washed before eating (you are the one who knows what kind of chemicals you've been putting on it), and, for my taste, it should be added to other plants and then covered with a good vinaigrette. How much different is lawn grass from the young wheat grass blades commonly added to today's health drinks?

DANDELION

The dandelion is one plant that will likely be in everyone's lawn. You work to get rid of it, but every breeze that blows across your yard spreads the

seeds, assuring survival of the plant. The dandelion is rich in vitamin A and is loaded with other nutrients as well. It can be used in salads, cooked greens, and fritters. The mature green leaves are bitter (the sun and the chlorophyll work to make the green leaves bitter) and are best if boiled two or three times before consumption or slathered with a dressing. The best leaves are those you find growing under something (like a board) or otherwise not directly exposed to the sun. These leaves will be nearly white and much less bitter. The best part of the plant anytime of the year is the rosette, where the leaves have not yet seen the sun. The best salad is made of young leaves still very light green in color (again, no color is better). A gallon of apple cider vinegar should be included in every food storage plan to facilitate salads and processing dandelions.

The crown, or the area between the upper leaves and the roots, is usually not directly exposed to the sun and is mild and tasty. The root system is also usable, but not for salads or the pot. The root should be washed well and then roasted by putting it in an empty can and then putting the can on the coals in a fire. Turn the can often to keep the roots from burning. The roots will end up dry, and you can then crush them up for various uses, including as a coffee substitute (without the caffeine). The flower can also be dipped in batter and fried into a fritter.

PLANTAIN

The plantain is another plant likely growing in your lawn. It is easy to recognize by its deep green, broad leaves. The leaves are usable as a pot herb (a leafy herb that is cooked for use) or in a salad. The ribbing is made up of fibers that are a little tough and take a little getting used to.

PURSLANE

This plant is abundant in the average lawn, as well as your flower garden and along the edge of your sidewalks and driveway (often in the cement cracks). It is generally a prostrate plant with succulent green leaves and reddish stems. The leaves are paddle shaped. Purslane is rich in iron, vitamins A and C, and calcium phosphorus. The leaves give off a slimy, mucilaginous, gumbo-like residue when crushed or cooked, much like okra. The residue thickens soups, however, and is a valuable component. Purslane can be gathered in volume and grows back fast, meaning that it may be the most valuable plant to augment your food supply on a daily basis.

CHICKWEED

This plant is very abundant in your lawns, flowerbeds, and gardens. This is not well loved by the average homeowner, though, because it is so hard to get rid of. But once your need becomes survival, it is valuable as a food source. When you learn how to recognize it, you will have no trouble finding it. It can be used as either a pot herb or in a salad.

CLOVER

Clover is easily identified, known by most everyone, and is widely available. Clover leaves can be used as a pot herb or a salad and, overall, the plant is very wholesome and nutritious. The flower is the best part of the plant, so make sure you include it along with the rest of the plant.

SHEPHERD'S PURSE

This is a member of the mustard family and has the pleasing peppery taste that you would expect from mustard. The whole plant is edible and can be eaten raw, in a salad, or it can be cooked, salted, and served as a pot herb. It is not generally found in lawns, but it's common as a weed in most gardens.

LAMBSQUARTERS

This is perhaps the best-known member of the family Chenopodium. It's a common yard plant family that includes pigweed and amaranth. These plants are so much alike that they are often grouped under one name—lambsquarters. Both are edible and nutritious. They are better than spinach, in my opinion, and make a great salad or potherb. The seeds can be used as gruel or ground up and made into bread. Lambsquarters is usually found in your garden.

ROSE

Rose petals are good to eat and can be added to your salad. They add taste and color to food. However, in my opinion, the rose's greatest value is its rose hips, which can be used through the cold winter months. The hip is generally used with its flesh and seeds. It is used in making jams, jellies, and rose hip tea. There are high amounts of vitamin C valuable to stave off scurvy.

GROUNDCHERRY

This plant is recognized by its low-trailing vines with yellow flowers that later form little husks with small tomato-like berries inside. (To me, calling them ground tomatoes would make more sense because the berry is more like a very small tomato, both in shape and in flavor.) The berry can be eaten raw, added to salads, or cooked into stews or pies. Use the berry only after it is bright and yellow.

The plants listed above will likely be available to you if you've bugged out, but in more rural areas, other plants can be added to the menu. Maybe you've already purchased books or other articles to help you in that effort. The resources will likely be more comprehensive than what is described here. If you have already made that move, I congratulate you and suggest that you continue the education process.

ARROWHEAD

This plant is easy to recognize by its distinctive arrowhead-shaped leaves. It grows in ponds or slow-moving streams and is nutritious and delectable. It sends out runners a few feet from the plant, and that's where you will find the starchy root tubers. The best way to harvest them is by wading into the mud and shoving the mud around the base of the plant. You'll need to disturb the mud several feet around the plant fairly deep. Once you have dislodged the tubers, they will float to the surface. The tubers range in size from the size of small BB shot to large quail eggs. They can be eaten raw but taste better cooked.

ARROWLEAF BALSAMROOT

This is a shorter plant with arrowhead-shaped leaves. It has a yellow flower that looks like a sunflower. The leaves are covered with fine hairs. The leaves and the stems are edible. The large roots can be eaten. However, they are strong tasting and require boiling.

ASPARAGUS

No description should be necessary here because many of us know what this plant looks like. It can be found almost everywhere and is a plant that you should keep an eye out for. If you find a fairly abundant supply, use only the new shoots. It generally grows more abundantly along ditch banks or fencerows.

BISCUIT ROOT

This is a low-growing plant. It is a member of the carrot family. The plant tends to grow in arid places in the West. Some close relatives are poisonous, so you will need to learn to recognize the edible ones. It will be well worth the trouble, though, because the plant is a great source of food in places where food is rare. The flowers are found on compound umbels, and the leaves are like those of a carrot but low to the ground.

The large root is great in stews or flattened into cakes and cooked on hot rocks.

BRACKEN FERN

One of the most common wild plant foods is fern. The only edible part of fern, however, is the young shoots. The shoots are called fiddleheads (because they look like the top of a violin) and they should be used as a pot herb.

BURDOCK

This plant looks like rhubarb without the red stalks. In addition, the leaves are smooth and velvety to the touch. The first year plants are the ones to look for because they are the most tender and tasty. The stalk should be peeled before eating, and the root should be washed and peeled. This plant is easy to recognize by its seed head, which consists of burrs with spines that stick into your clothes.

BULRUSH

This plant is tall, often as tall as a man. It grows in water, usually in lakes and swamps, and is found in large stands, thickly matted together. The part of the plant that is edible is the rootstalks and the new shoots before they break free of the water. Harvesting them requires walking out into the water. The best place to find shoots is along the outside edge of the stand, where these plants are trying to spread out. You need to thrust your hand deep into the mud near the outside plants and feel for the rootstalks; then move along it to locate the shoots. The seeds can also be ground and used as powder or gruel.

CAMAS

Any effort to find and harvest this plant will take a lot of study because there is another plant that looks like it but is very poisonous. It is called death camas.

The flower death camas is usually yellow-white, and the flower on the camas is anywhere between creamy white to bluish white. The leaves are too much alike to be of any help in identification. While camas grows in moist meadows, the death camas will generally be found in drier environments. Usually the bulb on the regular camas is larger. The only way I know to be sure is to have both plants in your hands at the same time. It's kind of like comparing a head of cabbage to a head of lettuce. However, it is worth getting to know the real camas because the camas bulb is one of the best wild plant bulbs out there. The bulb can be roasted on the coals of the fire and then either eaten or flattened into cakes for storage.

CATTAIL

It is a source of high-value food, especially during seasons when little else is available. Any time of the year, even in the dead of winter, the rootstalks can be eaten. In the spring, the young shoots are tender, and a little bit later in the season, the reproductive stem appears with its two-piece sausage-like head. The stem is used in two ways: when you beat the head over a container, you get golden pollen, which can be made into flour to make muffins, pancakes, or bread. Added to wheat flour, the pollen works as a good "stretcher" and adds a wonderful new flavor. The green head can be cooked and eaten like asparagus. Every part of the cattail is edible, and the dry leaves can be woven into baskets, sleeping mats, or shelter walls and roofs.

CHICKORY

This plant's leaves look like the leaves of a dandelion, but they are on a stem rather than a rosette. The plant is relatively tall, and it has blue flowers (rather than the yellow flowers of the dandelion). The leaves can be used as a potherb, but the real prize is the roots. If you are from the South, you probably already know

about this use, because it's very common to add to chicory to coffee for added flavor. The process used is to roast roots and crush them. The dry roasted roots crushed up and added to boiling water is your new coffee.

CHOKECHERRY

This is a tree, and the edible part is the cherries that form in clusters of dark red. The inside of the cherries have a substance in them called Cyanogenic acid, and I never eat the seeds. Others (like Native Americans) did and seemed not to suffer. For me, it is the flesh of the cherry that I value. It is astringent and takes getting used to, but it is a great source of sugar.

CURRANT

The look of this plant somewhere between a tree and a tall bush. It has many berries hanging in clusters. The berries are yellow, red, or sometimes nearly black. Berries are great right off of the bush, and they can be used fresh or dried. They make great jams and jellies and are one of the best berries found in the wild.

ELDERBERRY

This is a large shrub with brown stems and compound leaves. There are two main varieties. One has clusters of deep-purple berries; this is the best known and has the best flavor. The second type is red (some people think it is poison, though there is no evidence of serious poisoning from eating it). However, something in the berries can cause some to become nauseated after eating a few berries. Strangely, though, when added to a bunch of crushed up ants and a little sugar, the nausea seems to no longer to be a problem. There is no question that the purple variety is the best. Elderberries make good syrup, jams, jellies, and wine.

EVENING PRIMROSE

This is a low-growing plant. Its flower is generally white. It has long leaves coming from the crown of the root. The flowers remain open during the night. The seeds can be eaten parched and ground into meal. The roots are bitter but can be cooked twice in boiling water to make them palatable (be sure to discard water after the first cooking).

GOLDENROD

This is a tall plant with a cluster of yellow flowers. The seeds are usable as gruel or as a thickening agent for stews.

INDIAN POTATO

This is a tiny plant with basil-like leaves. These leaves resemble bird tracks. The stem grows from a deep bulb. This plant was a mainstay of Native Americans. When you are digging for the bulb, care should be taken not to break the long underground stems because it will be harder to find the bulb if the stem is broken. The bulb will be located a little deeper than you may expect. The bulb is very good raw, boiled, roasted, or steamed. If you have found a good patch, you may want to flatten the bulbs and dry them for later use.

MAPLE

You can eat the winged seeds. Separate the seed from its husk, and then boil the seed and use it in soups or stews. In the spring, you can drain the sap by cutting into the bark and extracting it for a sweet drink. If you can harvest enough, it can be cooked down into syrup.

MILKWEED

This is a very common plant. The flowers are arranged in a cluster. These can be dipped into a batter and deep-fried; the seedpods can be eaten while

they are young and small; and the fibers in the dried stalks make very strong cordage. The only caution is to make sure it's not the dogbane, which is poison, but also makes great cordage.

MALLOW

There are a number of different mallows. They're all edible. Mallow is informally (in some parts of the US) known as "cheesies" or "cheeses." The mallow family is large and usually identified by the seedpod, which is shaped like round blocks of cheese—hence the nickname. Children often pick the little round seedpods and eat them. While they don't have much flavor, they are fairly nutritious, easy to recognize, and all parts are edible. You can boil a handful of roots for about thirty minutes and then remove the roots, which are now ready to eat. The leftover liquid is also usable: just beat the liquid like you would egg whites until it reaches a meringue-like consistency.

MINER'S LETTUCE

This is a small, succulent plant with leaves that are somewhere between round and arrowhead shaped. This is a salad herb that is so good you would choose it over lettuce. To cook it would spoil it.

MINT

There are a lot of useful plants in this family. The family is characterized by having a square stem. Some of the members of this family are sage, spearmint, peppermint, horehound, health, and thyme. They are useful mostly as a drink (usually a tea).

MORMON TEA

This is a desert shrub with green-colored stems. There are no leaves. This plant is only good as a tea, but it's a good tea, and it tastes best with a little sugar, lemon, and ice.

MULE'S EAR

This plant that has a flower that looks much like a sunflower. The leaves are glossy. The seeds can be eaten after a long cooking.

MUSTARD

Mustard is a common plant. It is leafy with white or yellow flowers, and as you might guess, it tastes like mustard. It makes a great pot herb and is used as such throughout the South. Mustard greens are very popular in many parts of the country.

OAK

The edible part of the oak is the acorn. It has been used by Native Americans for centuries. The acorn contains tannins, however, and needs to be mashed and washed to remove the astringents in them.

WILD ONION

Wild onions resemble domestic onions but are smaller. Their leaves are grass-like. When you crush the leaves between your fingers, you get that familiar onion-garlic smell. Once you learn to recognize that inflorescence or flowering top, you can even harvest it in the winter. The whole plant is edible.

OREGON GRAPE

The wild Oregon grape is a holly-like plant usually growing in dense thickets high in the mountains. The deep-purple berries are edible and tasty.

PIÑON PINE

Unlike other pine, this plant grows more in the shape and size of a juniper tree. Often the types of tress can be found together. This is the best pine for procuring large nuts. Harvest time is late summer to early winter.

PRICKLY PEAR CACTUS

Most of the cacti found in this country is edible. The prickly pear is the most common, however, and is found in many parts of the country. This cactus has leaves that are shaped like a pear, only flat. Because they are covered with sharp spines, they require careful handling. The secret is fire. Place each leaf into a fire, and then quickly flip it over to burn off the spines on both sides of the prickly pear leaf. The easiest approach to eating it, after removing the spines, is to slice the leaf into strips and add it to your soups or stews. The flowering bulb at the top is also a valuable food. It is red and full of sugar. After you burn off the spines, you can eat the bulb like any fruit. You can also turn it into jam, jelly, or syrup.

RASPBERRY

The wild raspberry is almost identical to the domestic variety. The berries are the main part to harvest. However, the leaves can be made into a very good tea and are useful to calm morning sickness or general nausea.

SALSIFY

Sometimes known as oyster plant because its roots seem to have a slight oyster flavor, the second-year plant has a yellow or blue flower. However, it is the first year plant you are looking for, and it produces only leaves. Once you learn to recognize the plant, you will then be able to recognize the difference between the first- and second-year plants. This plant is worth learning how to identify because the steamed roots are delicious.

SAMPHIRE

This plan is also sometimes known as salicornia. It is a low-growing plant that is found in brackish, saline environments. It is one of the few plants that will grow in saltwater. It is all stems, no leaves, and the stems are fleshy and edible. They can be eaten freshly picked or added to soups and stews. The stems are salty, and because your body needs and demands a certain amount of salt, this plant is critical as one of the best sources of salt in the wild.

SERVICEBERRY

This plant is more of a tree than a bush, but it grows in thick stands and yields an apple-like berry. The berries are deep blue in color and can be eaten raw or cooked. Serviceberry is often used as an ingredient in pemmican and can also be used in pies.

SOUR DOCK

This plant and a close cousin named sorrel dock are two of the most commonly used dock plants. They both have large, dark-green lance-like leaves. The seed heads are brown in sour dock and reddish in sorrel dock. They both will serve well as pot herbs, salad ingredients, or raw food. The seed heads can also be used as emergency gruel.

SPRING BEAUTY

This is a tiny plant that appears in early spring (sometimes right after the snow has melted). On the surface, there will be only two leaves opposite each other on the stem. The flowers usually appear in clusters of either white or pink. You'll have to dig down several inches to find the edible bulb. It can be eaten raw or cooked like a potato. The bulb is usually about the size of a child's marble.

STRAWBERRY

If you have never tasted a wild strawberry, you haven't had the real strawberry experience. The plant is much the same as a domestic strawberry plant, but the berry is smaller and tastes better than a domestic strawberry.

SQUAWBUSH

Not to be confused with sumac, which has similar-looking berries, this is a fairly short bush. It tends to grow near streams or in damp environments. It is easy to recognize by the leaf, which grow in what appear to be three leaves on one stem. The berries grow in clusters, with each separate berry about the size of a BB. The berries are sticky but make a nice drink, which is often called "wilderness lemonade."

SUNFLOWER

Sunflowers are common everywhere. The seeds are very nutritious but are small and can require substantial effort to harvest.

THISTLE

There are many species of thistle. Generally, thistle can be found in open fields or in small wooded areas. Thistle is easily recognizable. (Though if you have doubts about your ability to recognize them, walking through a stand of the sharply pointed plants will help you decide.) The whole family is edible, but some are more palatable than others. The flower resembles a shaving brush. Only the base of the flower can be used, but it can be eaten like an artichoke. The most valuable part is the stalk—especially in the spring, when it is juicy. The outside of the stalk is, like the rest of the plant, covered with very sharp spines. It needs to be carefully skinned with a pocketknife. The end product is worth the effort, and the plant is like a more tasty and tender version of celery.

THORNAPPLE

Also known as haws. This is a tree with thorns with fruit that resemble miniature apple. The berries are usually red, unlike the serviceberry, which are blue. The berries can be eaten raw or baked into a pie.

VIOLET

(DOG'S TOOTH LEFT), SEGO LILY (RIGHT)

These are two of the many varieties of violets. The leaves can be used in salads, but the best part of the plant is the bulb. It is usually buried deep but is well worth the effort to dig up.

WATERCRESS

This is a green leafy plant that grows in small, clear streams. Watercress is a plant that will improve any salad. My favorite way to enjoy watercress is between two slices of bread with butter. It has a peppery taste. If the water where the watercress is growing is no longer safe to drink, then neither is watercress. The plant may include some harmful bugs. Wash it, and then soak it in water that you have added a couple of drops of iodine to.

WATERLEAF

This is a beautiful little plant that is found at higher altitudes. The leaves are bunched together with several leaves on each stem. The flowers grow in globes of white to light purple. The whole planet is edible and good. The rootstalks are like several tiny carrots blended together. They taste great in stews or eaten fresh.

WILD RICE

This is also known as Indian rice. It is recognized by a large plume on top of the grass. It grows most abundantly in the Eastern US. The plant grows in water with a mucky bottom and shallow water where there is enough circulation to prevent stagnation. It is easiest to harvest in the late summer or early fall, when the plume is getting ready to drop. Native Americans had a very effective way of harvesting this plant. They spread a large cloth across the bottom of a canoe and paddled among the plants, bending the stalks over the canoe, and beating the rice out of the plant with a paddle or stick.

YAMPA

This plant tends to grow in high meadows, hillsides, and foothills. The leaves are usually dry by the time it flowers. Its flowers are white or yellow. The small finger-like roots are generally buried deep but are worth the effort to dig up. They are better than potatoes in flavor.

There are many more edible plants you can use, but the plants provided here are less likely to be confused with toxic look-alikes. I leave it to you to learn to recognize others. Mushrooms have been left out for the same reason. While the majority of mushrooms are edible and delicious, there are a few that are deadly poisonous. The only safe approach is to learn each edible mushroom and use only those that you can identify. Buy books dedicated to botany and edible plants and learn the plants common in the area where you live. Having knowledge about edible plant use will be an invaluable asset and will prevent you from starving while sitting in a field of food.

NOTES

1. Bradford Angier, *Field Guide to Edible Wild Plants: a quick, all-in-color identifier of more than 100 edible wild plants growing free in the United States and Canada* (Harrisburg, PA: Stackpole Books, 1974), 22.

13 FRESH ANIMAL FOOD

> "A good rule is not to pass up any reasonable food sources if we are ever in need. There are many dead men who, through ignorance or fastidiousness, did."[1]
>
> –Bradford Angier

Finding fresh meat will be more difficult for those who bug in than it will be for those who bug out. You will be left to make do with the animals in your neighborhood. How many of us commonly have deer in our backyards at night? Maybe a few people might have had that experience but not many. The majority of animals moving around in your neighborhood will instead be dogs and cats. These will be domestic animals, either belonging to you or to one of your neighbors, though it is likely that many of these animals will have been turned loose by their owners because they could no longer afford to feed them. It has been suggested that domestic animals gone feral, either from being let loose or of their own choice, will be a plague to be dealt with during TEOTWAWKI. Packs of feral dogs running wild, with no food, will be commonplace. These packs will be made up of big dogs, too, because the little dogs likely won't survive encounters with the hungry big ones. It won't take long until it will no longer be safe to venture outside. Small children will become the dog pack's favorite targets. Some kind of effort will need to be taken to solve this problem—and as soon as that begins, someone's going to figure out that there is fresh protein to be had. The thought of consuming a dead domestic pet is likely repulsive for most of you and is probably just a step above cannibalism. However, hard times call for hard decisions, and many will likely overcome this distaste of eating domestic pets.

There are a small number of fresh-meat options for those bugging in. For example, rats are common in many cities and could make a valuable addition to menus during TEOTWAWKI. Birds, too, are common, and several species are large enough to be worth eating. The mourning dove and Eurasian dove, both hunted for their meat, are common visitors. The domestic pigeon will make good eating also. Ducks, sometimes up to a hundred at a time, are common in some urban and rural areas, as are droves of California quail. Other flock birds, like starlings and blackbirds, are also frequent visitors to most of our yards. You need to evaluate the situation around your house, because initially, only those wild resources in and around your house will be available. Most birds will be too small to satisfy the appetite of one person. So the best strategy would be to include the meat of whatever you have harvested to soups, stews, or casseroles. However, those of us who have chosen to stay put will generally not have much fresh animal protein to count on. For those who have chosen the bug-out option, though, the possibilities are better. When you're in the woods, you will find more animals to hunt and harvest.

Some preppers, however, don't feel the need for animal protein. You have been getting by on plant foods just fine for years. Let's examine a condition commonly known as "Rabbit Starvation." Let's imagine a situation where you and your family are forced to land a small plane somewhere in the wilderness. The engine will not start, and the radio is inoperable. The good news is you have a couple of .22 rifles and plenty of ammunition. You also see that wild rabbits are everywhere. You are not going to starve. On the first day, you bag a dozen large rabbits in less than an hour. During the first week, you have no trouble bagging more rabbits than you can eat. Because you have no cooking utensils, you are forced to cook the dead rabbits on a stick, like you would cook marshmallows over a campfire. But by the end of the first week, you and every member of your family have come down with a mysterious disease—you all have a serious case of diarrhea. Each day it gets worse until, one day, your youngest daughter dies. By now, you are all too weak to help one another. You are all too sick to eat, but you all continue to get sicker. By the end of the second week, there is no one alive to rescue.

It is fairly common knowledge that a man can survive at least thirty days without food, so what happened? Death by rabbit starvation is based on a lack of fat in your diet. Your intestinal tract must have a certain amount of fat in your diet. Rabbits store their fat in their intestines. The end result is no matter how much rabbit meat you eat without fat, you will die faster than if you went without food. It does not matter how bad your dietitian makes fat sound, you must have some of it in your diet or serious diarrhea will be the result and you will die. Several different types of oil can be added to your food storage to help. A good oil to add is EVO—better known as extra virgin olive oil. You might follow that with some butter (food storage stores carry butter in cans for long-term storage). We need to be careful not to exclude the essential fats. There is a reason we refer to them as essential. Many diseases are caused by deficiency of certain fats, and many of the same diseases can be overcome by supplying the essential fats missing from your diet. First are the omega fats: omega-3, omega-6, and omega-9. The omega-3s help us avoid rheumatoid arthritis and other inflammatory diseases. They also help with the normal blood flow in your arteries. Omega-6 is found in sunflower and corn oil. Omega-9 is normally found in olives, almonds, macadamias, hazelnuts, peanuts, sesame seeds, and avocados.

You need to understand that in a survival situation, everything will change, and nothing will change more than how you eat. Built in to many of the foods we consume on a daily basis are things like fats. Bread has a certain amount of oil in every recipe. Look at the ingredients on manufactured foods. They all contain fat. According to Dr. Atkins, "Many diseases are caused by deficiencies of certain fats."[2] Our storage should contain things like vegetable oil, olive oil, sunflower oil, peanut oil, and canola oil. And most important, we need to include plenty of fresh animal fats in our diet. In the imaginary plane crash, the rabbit meat would have saved the family if they had included the fat stored in the rabbit's intestines by cooking the intestines. The fat in the broth would be just about the right amount to keep away the diarrhea.

To some, the idea of eating fat is distasteful. But you do it all the time without thinking about it. The real reason we pay more for a steak at a high-end steakhouse is because their steaks taste better because their meat buyers by the highest quality meat, which is the meat with the most fat marbling.

If you have type-O blood, you are called "the Hunter" by Dr. Peter D'Adamo.[3] According to him, your success depends on the use of meat poultry and fish. If you are a type A, you will do better with plenty of vegetables. But no matter what your blood type is, you'll need some meat in your diet. Several sources for animal protein are described below.

SMALL RODENTS

This includes animals such as moles, mice, or lemmings. The only sensible way to prepare something so small is handling them like the early Native Americans did—cooking them whole or as part of a stew. You can discard the entrails after the cooking process is over, because most of the high-value fats would have cooked into the stew.

MEDIUM-SIZED RODENTS

This includes rats, rabbits, gophers, or squirrels. These can also be cooked like small rodents, but because they are larger, a little more care could be used. I have friends who live out in the swamp, west of Manny, Louisiana. I remember the first time I ate with them. I was served a platter of rice and okra with fried squirrel served on top. But the guests were also served a squirrel head on top of the pile. I wasn't sure what I was supposed to do with it, so I watched the others. As soon as I saw them crack into the skull, I also picked out the brains and swallowed them down. It was not a big deal to me because my grandpa had been serving sheep brains and scrambled eggs since I was a small boy.

One note on eating the brain of an animal, there are some diseases—such as chronic wasting disease—that are thought to be contracted by eating the organs of diseased animals like deer and elk, including their brain. The CDC has issued the following statement: "As a precaution, hunters should avoid eating deer and elk tissues known to harbor the CWD [chronic wasting disease] agent (e.g., brain, spinal cord, eyes, spleen, tonsils, lymph nodes) from areas where CWD has been identified."[4]

LARGER RODENTS:

Large rodents include skunk, muskrat, and opossum. These are often cooked by browning the pieces on all sides in hot fat and adding seasoning to taste.

Muskrats are water-loving animals found in and around ponds, streams, and rivers. My oldest son has been a trapper for many years, so I have enjoyed a lot of muskrat meat. One of our favorite methods for cooking muskrat is popping the skinned animal into a smoker. At that time, we were living in a small community called Dutch John, Utah. When we'd cook muskrat, a large number of the town's residents would line up for a serving.

The ubiquitous skunk also got caught in many of my son's traps. Care is needed when removing them from the trap, and even greater care is needed when eating them. Once you manage to get around the scent glands, the skunk can be smoked just as well as muskrat.

I have also deep-fried woodchucks. Opossums are also palatable. My friends in western Louisiana and east Texas introduced me to opossum. I found it to be a little oily, but in a survival situation, that's a good thing. The most common way they are cooked is by removing the glands at the small of the back and inside the four legs and then preparing it like you would a woodchuck.

ARMADILLO

For a short time, I was stationed at Fort Hood, Texas. We were preparing for a large training operation. Consequently, my outfit spent all of our time out in the field. For several months, I never saw the inside of the barracks; we were always moving too fast for the mess tent to catch up. So we lived on K rations (the army's field food before MREs). There may be worse foods than K rations, but not many. I had a couple of local boys in my squad, and they introduced me to armadillos. We cooked them in our steel helmets.

RACCOON

When I was a small boy, the only thing I knew about raccoons was what I read in books. There were no raccoons where I lived. However, raccoons have now moved into the area, and there are so many that my farmer friends beg me to come and trap or kill them. But they are so abundant that any effort to eradicate them would be like trying to exterminate a whole ant bed one ant at a time. Over the years, I have trapped enough raccoons to enjoy a good number of raccoon dinners. They are good eating. I generally use the same approach as with woodchucks. Skin and cut them up into pieces, parboil them, roll them in flour, and cook them like southern fried chicken.

You should not waste the skin. Davy Crockett didn't invent the coonskin cap; the Indians were using them as headwear long before the pilgrims stepped off the Mayflower. When you begin living in the woods, you need to stay warm in the winter. A good raccoon hat would provide warmth.

PORCUPINE

This animal is a walking bundle of needles, and it may not look like food, but it is. The flesh is good cooked any way you like, and don't throw out the liver—it's very large considering the animal's size, and it is mild and tasty. Preparing the animal is a challenge and is best done very carefully. Roll the animal onto its back after making the first incision, then roll the skin and use that inside part to hold it as you finish skinning the rest. The porcupine can be easily killed without a gun. When I'm hunting them, I usually carry a special walking stick. It is about six feet long and a bit thicker around than my thumb. Porcupines have concave spot in their skull even with its two nostrils, and a good whack with your stick at that point will drive a bone into the animal's brain. It will roll over and be ready for you to work on.

BEAVER

A beaver is a relatively large animal and will feed a whole family—with leftovers. Its pelt is good for winter clothing. The meat is wonderful and can be cooked in a dozen different ways. The tissue in the tail is an especially important piece of the animal. It's not really fat but is something similar to it, and it's firm, rich, and extremely nutritious. To cook a beaver tail, hold it close the fire and wait until it's black and covered in blisters, then flip it over and cook the other side the same way. The skin will then easily peel off.

TURTLES

In the continental US, the only turtle worth the time and effort to gather as a meal is the snapping turtle. It should be handled carefully. Keep your fingers away from its mouth. After gathering and killing the animal, start preparation by removing the bottom or belly plate. This is done by cutting through the sections in the middle and by each leg. Carefully remove all the meat you can, and then soak it in salt water for a few hours; you can either freeze it or parboil it to tenderize it. Then it can be fried or made into turtle soup.

SNAKES AND LIZARDS

In the US, most lizards are edible, but some are simply too small to be worth the effort to catch. One lizard, the gila monster, is venomous and not worth the risk to catch. (Though if you are near starvation, trying to capture a gila monster might be worth it.) The trick to capturing a gila monster or any snake is removing the head without getting bit. You can hold the head down with a long stick. While you are holding the head, down carefully cut the head off with a knife. After the head is removed from the body, kick it away from you with your foot. Don't try to pick the head up. Complicating matters somewhat, some of the best food-source snakes are the most poisonous ones. The reason for this is pretty simple: venomous snakes use their muscles only when moving, while constricting snakes use their muscles for crushing the life out of their prey. Consequently, the meat of the constrictors is a lot tougher. During my years in the jungles of Central and South America, I learned a good trick for getting around this problem. I found if I wrapped a four-inch piece of constricting snake in the leaves of a papaya tree for a few hours, the meat became very tender. Though you may not be able to find a papaya tree growing in your neighborhood, you can still employ this method. The thing that makes any meat tender is an enzyme called papain, which is the main ingredient in most meat tenderizers on the market. If you add a few bottles to your food storage, you will be able to handle any tough meat without a lot of parboiling.

I have not heard of any snake meat that is not edible. Many years ago, I ran a survival trip for a small college in eastern Oregon. It was late fall and the area had been in a serious drought for several years. The mountains were dry and without anything we could survive on. We quickly found out that the only thing available to eat—and it was available in big numbers—was rattlesnake. We had very little trouble finding them under every rock and log. This is the method I taught my students: each of them selected a willow stick, which was cut into six-foot lengths with a small fork at the end. The fork had to be strong, with only an inch on either side. Any longer and the snake could pull itself free, often just when you didn't want it to! The larger end of the willow was used to move the snake out of its coil (because that is when it can strike and is the most dangerous and hardest to get at). The key maneuver is to get the fork just behind the head to hold it down; you then cut off its head with a knife.

Just because you have managed to cut off its head doesn't mean it is now safe to work on the snake. You will need to get the dangerous part, the head, away from the area before you can begin to process the snake. Do not try to pick up the head with your fingers. Use your stick to move it away. (This step cannot be taken lightly.) At the upper end of the snake, peel the skin off, and then move down the snake, peeling as you go—like removing a sock. When you get to the rattle, lay out the skinned meat and cut it into separate pieces. Once the skin is removed, the guts will come off. The meat can then be cooked.

BIRDS

All birds are edible. However, some of my fishermen friends have told me that it is not wise to eat seagulls, because they can make you sick. I don't think it's the bird but rather what they've been eating. It's not uncommon to have small, localized clusters of dinoflagellates, commonly known as red tide, that give off deadly toxins that kill everything in their area. Seagulls are not very discriminating eaters, so when they see dead or dying fish, they rush to feed. The birds not killed will still have enough toxins in their tissue to make you sick. So when you are near the sea, it may be wise to scratch seagulls off your edible animals list. All other birds are fair game. Most small birds—like sparrows or robins—are too small to feed a family. They will, however, serve as a meal for the hunter. Generally, there is very little in the way of meat anywhere other than the breast. Still, small birds can be a great source of protein. The game birds, such as doves, quail, pheasant, and grouse, are readily available and provide a larger source of protein. Some species, like quail and doves, also occur in urban and rural environments. Waterfowl are also common in urban and rural settings and include numerous species of ducks, geese, and swans. Most of them are

large enough to serve as the main dish of any meal. Wild turkey is also plentiful in some areas and will provide a large meal.

PREDATORS

This category includes foxes, coyotes, bobcats, mountain lions, wolves, and others. They may not sound like food, but I have eaten them all and can attest that they are edible. Remember, in a survival situation no food should be turned down. They can also be turned into clothing, which is especially valuable during cold months. If you are of bug-outer, you should have included some good traps in your equipment, which will come in handy when pursuing predators.

WILD PIGS

There are three different varieties of pigs in America. The first is the "almost wild pig," or semi-feral pig. It is a pig like you would see on any farm, but it has escaped or has been turned loose by the farmer because the price of the meat fell below the cost of feeding it. It might be spotted or even have white hairs.

The second kind of wild pig I will call the "mostly wild pig." This is the descendent of pigs that have been wild for many generations and has begun to take on the appearance of its wild ancestor.

Then there is the wild pig. This is the same as the wild European or Russian boar. It has very long, dark brown or black hair and is built front heavy. The males have long, wicked tusks that are not something you want to mess with. All three types of pig can be aggressive and dangerous to hunt, so take care. All three are good eating, but the older they are, the tougher the meat will be. A younger pig is not only better eating but is also easier to handle. There are wild pigs in almost every state in America. In most states, they have become so numerous that land managers would like them eradicated.

If you happen to live in a southwestern state, there is another pig too—the wild javelina or peccary. It is a much smaller pig but is very tasty as long as you dress it the right away. You need to cool the meat quickly and remove the scent gland immediately.

BIG GAME ANIMALS

This includes deer, elk, antelope, caribou, and moose. I choose to spend the largest amount of time on deer, because it will likely be readily available to you. There are three main types of deer in the US. The whitetail is the most common. You can find it in almost every state in the union. The next numerous is the mule deer, with the black tail falling in last. The flavor of the meat is the same, with subtle differences depending on what each animal is feeding on. The quality of the meat is also influenced by the way the animal is handled in the field. While working as a law enforcement ranger, I checked hunters during the season and was often shocked to see dead deer in the back of pickup beds that had been there all day without being skinned or gutted. I couldn't imagine what the meat on those deer would taste like.

There are several key things that need to be done if your venison is going to taste good. First, bleed the animal immediately, and then remove the guts. Do whatever you can to cool the meat quickly. You may have to leave the hide on the animal until you get back to camp, but get it off as soon as possible. It is wonderful meat if handled properly, but hold off on one of these steps, even for short while, and the meat will quickly sour. Take a trip to your meat packing plant, and you will see them handle the meat as quickly as possible. Elk, caribou, and especially moose are much bigger and will spoil if you fail to cool them quickly enough. If snow is present, shovel some into the body cavity to cool the body.

High country animals such as bighorn sheep and mountain goats spend most of their lives in the mountainous high country. Because of this, most of us will not be able to harvest them, which is too bad because mountain sheep are one of the tastiest meats to be had.

Bears, too, can be a useful food source when needed. In America, we have two varieties—the black bear, which has both black and brown phases, and the grizzly bear. Both provide wonderful pelts that are exceedingly warm in the winter and make a good bed covering. Bear fat is a useful part of the animal too. On the black bear, fat is much like lard, and it is excellent to use in cooking. The fat on the grizzly bear is also good. For some reason, the meat of the black bear is favored by most and is much better eating then you might suppose. The meat on the grizzly bear is more gamy and is often passed over.

HUNTING FOR FOOD

You will need to know how to preserve meat—at least that meat coming from animals too large to consume in one meal. The first method is by creating jerky. Jerky is nothing more than meat that has been dried. You can add to its quality by smoking it and using salt, sugar, or a number of other spices to improve it. The meat should be cut into thin strips, dried in the sun or over a fire, and then salted and seasoned. For many centuries, people have also made sausage to preserve meat. Several books are available detailing how to prepare jerky, sausage, pemmican, and other dried meats. I recommend adding several to your library.

Many of you reading this book will think you have been hunting your whole life and don't need to be told how to do it. That might be true, but there are things that will be different that you may not have considered fully. For example, with all manufacturing ceased, you will no longer be able to buy guns or ammunition. Indeed, in a world where chaos is the norm, you might find yourself cut off from your home with nothing but what you have in your pants pocket. Things might be too dangerous to want to draw attention to yourself by firing a gun. In today's world, most hunters have available to them other tools that might be unavailable in the world after SHTF. You can forget about your four-wheeler, dirt bike, or pickup truck. That will change the way you hunt a lot more than you think. Forget about hunting guides or private hunting units. You will be on your own, and you may not be able to get to areas where there are game animals.

If you are fortunate enough to live somewhere where game animals are available to you, you may have to contend with more people who are now hunting the same animals in that area. The competition will be more intense than you would have ever expected, and the number of big-game animals will be declining every day. In today's world, very few hunters managed to fill a permit on a regular basis. How much harder will bagging a big game animal be in a chaotic world with no game limits or game warden to enforce order? So the first rule in the new world is "do not fail to harvest any source of animal protein that, at least in the past, you might have passed over." If you see a squirrel, you bag it; if you see a porcupine, you bag it. Whatever you come across is meat, so you bag it. You will no longer be free to be picky. You can no longer pass anything by.

Without question, the best weapon to hunt and harvest game would be a rifle or other gun. But in this new world, other people are very likely to pose a danger to you. You don't want them to know where you are. Shooting game with a gun will make a sound that will be heard for miles, and that sound may be a risk you will not be able to take. I suggest that you purchase an alternate weapon, like a crossbow or composite bow, and develop the skill needed to use it effectively. Another consideration would be your ability to harvest several different animals on each outing. A gun makes so much noise, it scares everything else away, but a good bow will allow you to continue hunting because it doesn't scare other animals away.

We need to be aware of the things that will change in the world after the SHTF. What should we expect to change, and what do we need to do to make ourselves ready to live in such a world? Preparation is the key to survival.

A WORD ABOUT YOUR PETS

Your food supply is designed to feed you and your family, but what about your pets? You probably think of them as family, but storing enough food for your real family is a costly business; can you afford to use it on your pets? Even if you could, they don't do as well, health-wise, as we do on such things as wheat and beans. So, the logical thing to do would be to also store a year's supply of dog or cat food.

During famine, you may feel the time has come for you to make the "hard choice," which is to turn them loose. If you make the decision to turn them out on their own, be aware that they are not prepared to live on their own. Most pets, as domesticated animals, are weak and unable to live without humans. This is mostly true of small dogs. In time, big dogs will eat them. Cats are never truly domesticated, so they can generally go back to living wild. However, the real question is, by turning them loose what have you created? Dogs will likely merge into hunting packs. No matter how sweet and kind they used to be, once they become part of a pack, they are dangerous. Man used to fear wolf packs, but we now know through biological studies in places like Yellowstone National Park, that wolves keep their numbers in check by eliminating rival packs when possible. Each pack will try to eliminate any other pack because of food competition. We don't know yet what dogs gone wild would do, but they may be less inclined to behave with that kind of efficiency. If that is the case, by turning them loose you have created another enemy. Unlike wolves, domestic dogs made feral are not skilled killers, so they will be in a constant state of hunger, desperate for something to eat. Domestic animals gone wild also do not have the same level of fear of humans. It is my guess then, that in time, humans will see the need for exterminating this threat. So in a way, by being unwilling to take the responsibility yourself, you have passed the task onto someone else.

In this day, almost every household has at least one dog or cat. If the city you live in is a large city with hundreds of thousands of inhabitants, then what we are talking about is a sudden release of nearly that amount of feral animals. That release would become a serious problem very quickly. Take a minute to think about it. Large packs of wild, starving dogs could become a problem as serious as small packs of starving men.

Think carefully about how you plan to handle the situation with your pets if the need should come to let them go.

NOTES

1. Bradford Angier, *How to Stay Alive in the Woods: A Complete Guide to Food, Shelter and Self-Preservation Anywhere* (New York: 2001, Black Dog & Leventhal Publishers), 24.
2. Robert C. Atkins, *Dr. Atkins' Vita-Nutrient Solution: Nature's Answer to Drugs*, 1st ed. (New York: Fireside, 1999), 196–221.
3. Peter D'Adamo and Catherine Whitney, *Eat Right 4 Your Type: the individualized diet solution to staying healthy, living longer & achieving your ideal weight* (New York: G. P. Putnam's Sons, 1996).
4. E. D. Belay et al., "Chronic Wasting Disease and Potential Transmission to Humans," *Emerging Infectious Diseases* 10, no. 6 (June 2004): 977–984, doi:10.3201/eid0905.020577.

14 FRESH FISH FOOD

"Fish is one of the healthiest foods in our diet. It is packed with protein, low in fat, and contains many essential vitamins."

–Unknown

For those who have decided to stay put, the chance to gather fresh fish is low. Unless you live right next to a river or stream, ranging afield to find fish would likely be a risky endeavor. Some bug-in people have prepared for this by creating a pond in their yard and stocking it with fish; others have built fish tanks in their greenhouses. These are wonderful ideas, because you not only get fresh fish for the table, but you also have a small water supply in the tank. The fish also fertilize the water in the tank, so when that water is used to water the plants in your garden or greenhouse, the plants are fertilized and grow faster and greener. You can also store canned fish.

Options for gathering fish are much better for the bug-out crowd. You should have chosen your bug-out location near fishable water. In addition, a large number of bug-outers will have chosen to locate near saltwater. A lot of you are planning your bug out under false ideas, though, and think fishing is an easy, leisurely activity. Most think that bringing home a supper of trout or salmon or largemouth bass will probably be a daily occurrence. However, that's what everyone else is thinking, meaning that you will experience heavy competition for fish resources. High-quality fish like salmon or trout will be on everyone's list. You might think that these are the only good fish and everything else is trash.

There probably isn't an angler alive who hasn't cussed a blue streak because he kept catching "them darn carp" or "them darn bullhead catfish," or whatever species it was that kept getting in the way of the prized game fish. Sport fisherman concentrate on a few species of game fish, but the so-called trash, or "rough fish," outnumber the game fish many times over. This can be turned to your advantage by changing your target and concentrating your energy on high-population trash fish. Many times, these regional preferences have to do with the fight of the fish type rather than flavor. For example, in Utah, perch are thought of as trash fish, but in other states, perch is rated higher because of its excellent flavor.

In the survival context, we are concerned with food more than the sport and suggest that for feeding your family, rough fish offer better odds. Your success with fish will depend on your location and proximity to water. People living in urban settings may have fewer chances of finding fish near your home, but some species of fish are nearly ubiquitous and can be found in drainage ditches, canals, erosion control ponds, and other similar bodies of water common in urban settings. Those living in more rural settings will have greater fishing opportunities.

During my first experience fishing for salmon in Alaska, I caught several fish called humpback salmon. My Alaskan fishing buddies told me I should just throw those back. "We don't eat them," they said. It didn't take me too much time to learn that in Alaska, many residents fish only for the "glamour" fish like the reds (also called the sockeye), or coho (called silvers), and the king salmon (known as Chinook). The other salmon are not considered worth the effort. While most Alaskans recognize that all salmon are edible, why eat anything but the best if you don't have to? In a survival situation, though, you will be glad to eat anything you can catch, and taste will become less important. Remember the special survival rule "no edible nutrition should be passed up." Most rough fish are tasty when cooked right. Sometimes you need to experiment to come up with a good recipe. Some fish are simply not edible until you learn how to properly prepare them.

TEOTWAWKI is not the time to pass up anything edible. It is a case of knowing what you are after and where to look. Some fish species, like salmon, can be found in fresh and salt water. Several prominent species are outlined in the following paragraphs.

SALMON

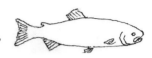

For the most part, salmon are not a freshwater fish. However, salmon come into freshwater to spawn, and this is when you will be most likely to come in contact with them. Number one on most fishermens' list is the red, or sockeye salmon. Like the name implies, its flesh is red during the spawning season. It actually returns to a blue hue while it is in the ocean.

Next on most lists is the king salmon, followed closely by the Cohos, or silvers. The humpback is next in popularity followed by the dog salmon. These five are all in the genus *Oncorhynchus*. There is another salmon belonging to a totally different genus—the Atlantic salmon, which has flesh similar to that of the silver salmon. If you live on the East Coast, this is the salmon that you will encounter.

TROUT

The most popular trout in most regions is the rainbow trout. It is a good-sized, tasty fish with pink flesh. The next most common variety is the cutthroat trout. The color of the flesh and flavor is much like the rainbow trout. The next is known as a salmon, but it is actually a sea run rainbow, called a steelhead salmon. The last is the brown trout (which is a European trout and not closely related to the other trout). It's flesh is also pink, and the flavor is good.

CHARR

Best known among the charrs is the brook trout (which should probably be called the brook charr). Generally, brook trout are a smaller fish, rarely over four pounds, but are very good eating. Next in terms of popularity is the lake trout (also a charr). It is much larger and, in some lakes, can weight over fifty pounds. The flesh is lighter in color but great tasting. It is fairly oily, but in a survival situation, that is a good trait. There is also a lake trout hybrid (called a splake) in many bodies of water. The lake trout spends most of its time in deep water and is rarely accessible to a shore-based angler, but the splake comes into water near the shore and can be caught without a boat or other deep-water equipment. The Dolly Varden trout, bull trout (also a charr), and Arctic charr are also reasonably numerous in various parts of the country.

BASS

Bass—including large and smallmouth—are one of the top glamour boys in sport fishing. They are fun to fish and are tasty to eat. Bass, unlike trout, who store fat in their flesh, store fat in their intestines, which gives their flesh a mild flavor. When you're ready to cook it, remove the skin, filet, fry, and serve.

SUNFISH

Sunfish are part of the bass family, and these smaller, round-bodied fish have the same kind of lean white meat found in bass. This should be relatively easy fish to procure while living in the wilderness. Bluegill is the best known of the Sunfish family, but there are other Sunfish large enough to eat. Too many, in fact, to name, and they are all good eating as long as they are big enough to cook. White crappie and the black crappie are also common in many regions, and they're generally caught in moderately shallow water and are cooked much like the rest of the Sunfish family.

CATFISH

The channel cat is, without doubt, the most popular of the catfish family. It has a body that is large and slender. The flesh ranges between lean and fat and is very tasty. Channel cats run a bit larger than most catfish and range between one to fifty pounds. Skin them first, and then filet them. Many sport fisherman don't like to fish for catfish because they're too "primitive." You

would be wrong to make the same mistake, because catfish are actually highly advanced. Their hearing is excellent, but their ability to smell and taste is their biggest strength. Not only can they smell food from a long way off, but they can also taste it at the same time. We think of these two things as separate, but when it comes to catfish, they do both at the same time. Their whole body is covered with glands that taste and smell. A catfish doesn't need to mouth your bait to taste it, he has already tasted and smelled it, and it was the taste of your bait that helped him make up his mind to take it. If you have interest in raising fish in a tank, the channels are best when not much over a pound. Other catfish include the blue cat, which is much like the channel cat but larger, followed by the flathead, which is also a big catfish. Each species can grow to over one hundred pounds. The smaller bullhead catfish rarely weighs over a pound but is good eating and much easier to catch.

PERCH

The perch is the one of the most heavily fished species—more for than any other member of the family. The walleye is the largest member of the perch family and the most glamorous. The walleye  has white, flaky flesh and can grow big for a freshwater fish—sometimes larger than twenty-five pounds. Its close cousin, the sauger is not as big but is just as tasty. The yellow perch is not nearly as big, usually only six to eight inches long, but it sometimes weighs over two pounds. What they lack in size, they'll make up for in flavor—small, maybe, but worth the effort. Perch run in large schools, and in a survival situation, perch are often relatively easy to catch.

TRUE BASS

This family includes the big striped bass, more commonly known as the striper. It is a big fish and

can reach over 125 pounds. It is a saltwater fish that has been introduced into many freshwater lakes. Another member of this family is the white bass, which is in fact a cousin to the striped bass that has been landlocked so long it has developed into a new fish. It is much smaller, usually four to fifteen inches long. It has been planted widely throughout the nation and is an excellent white-meat fish. The yellow bass, which is similar to the white bass in size and flavor, is common in the Southern US and in the Mississippi River drainage. This family also includes the white perch, which is largely similar in size, shape, and flavor to the white bass. The white perch is commonly found near the sea in brackish water and freshwater river mouths.

WHITEFISH

This is a relatively large family of fish, which includes the Inconnu of the North and the much smaller Rocky Mountain whitefish. Whitefish prefer moving water such as rivers and streams. They are bony but are excellent food fish.

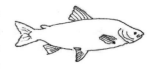

MINNOW FAMILY

Most people think of a minnows as any small fish. But the truth is, it is a large family of fish, and some species grow quite large. The Colorado River pikeminnow (formerly the Colorado squawfish), for example, grows very large. It has been reported to attain a length of six feet and a weight of one hundred pounds or more. I recall an old photograph of my father with one of these fish hanging from a pitchfork over his shoulder. My father was six-feet four inches tall, and the fish's tail touched the ground.

The common carp—the king of all trash fish—is also a minnow. A number of scientists call it the European carp because it was transplanted into the US from Europe, but even that is incorrect because it was transplanted into Europe from Asia. Why so much transplanting? The main reason is the fish has so much value as a food fish. The carp has been used and raised for many centuries as food. As a youth, one of my biggest pastimes was hunting carp with a bow and arrow or a pitchfork. My friends and I, I am ashamed to admit, wasted our catches because we didn't think the carp was an edible fish. As I grew older and wiser, I was never guilty of such waste. Very few fish take to a smoker better than a carp. And if you can't wait and want to cook it right away, soak the filets in milk and some other base, like soda, for a few hours before you cook it. This will get rid of the muddy content in the fish and the muddy taste. When taken in cold, clear water, you can pass up this process.

There are many smaller minnows that are still edible, though many of these are too small to make a good meal. Small minnows (and other small fish) can be captured and used in bulk, which improves their utility. If you want to use smaller minnows, they can be dried in the sun with a little salt and included in soups or fish stews.

SUCKER FAMILY

This is another large family of fish, all of which are edible. The flesh is lean and firm with a very mild, almost sweet flavor. They are very bony, however, which discourages many people from eating them. Suckers contain a small row of forked bones that run along the length of the fish just above the ribs, and when you locate this bone row, cut them away and you will end up with two narrow filets of delicious meat. Another way to handle sucker meat is by bottling it. The process will reduce and soften the bones to the consistency you see in canned salmon.

AMERICAN EEL

This is the only true eel found in America. Don't be fooled by the name. The American eel is a fish and

has a fishlike head, mouth, and gills. Eels are often confused with the lamprey, but the lamprey is more primitive and not related to the true eels. Unlike eels, they do not have a fishlike mouth. Rather, their mouth is something like a toilet plunger with teeth. It is round and filled with many hooked teeth. It attaches to other fish as a parasite (rasping away the flesh and sucking out blood). The true eel spends much of its life in rivers. It spawns in the sea, and the young eels then enter freshwater rivers, where they spend the rest of their lives until they too are ready to spawn, when they return to the sea. The American eel is common only in Atlantic Coastal waters. It has an odor unlike that of most fish and a flavor that can only be described as eel-like. The flesh of an eel is good and cooks well in stews. The hard part of preparing them is removing their skin, because they are very slimy and hard to hold. The best method is to use a piece of rough cloth to hold them while you cut through the skin all the way around just behind the gills. Once that is done, hold it with a cloth around its head while you work the skin down with a pair of pliers until it comes off. I like to cut the eel into two-inch pieces and place them in the pot for a delicious eel stew.

FRESHWATER DRUM

This is a member of the Croaker family found in freshwater. This fish can reach a weight of more than fifty pounds. While it is not the best eating fish, it is a large fish and is worth your effort. Because it usually tops two or three pounds, it makes up for lack of taste with quantity. It is mostly a bottom-feeding fish, which you will catch only if you fish with a bait on the bottom.

BURBOT

This fish is the only member of the codfish family living in freshwater. Where I live, it is called ling. It is

a fairly large fish, usually reaching a length of two or three feet, and it tends to come up to the shoreline at night to feed, when it can be caught in numbers. You shouldn't worry about overharvesting—this fish is numerous, is an extremely voracious feeder, and can seriously reduce the numbers of more popular species. This fish can be dried and preserved for future use. It can be bottled, smoked, or salted and dried. It may be an ugly fish, but it is good eating. It has lean, firm flesh and has a mild flavor.

PRIMITIVE GIANTS

In the US, we have four fish that fit the description of primitives.

The first is the sturgeon. There are several on both sides of the country; the biggest is the Pacific sturgeon, which can reach a length of over twelve feet and a weight of one thousand pounds. It is found from Alaska to California. On the eastern side of the country, we find the common sturgeon, which ranges from Florida to Cape Cod in rivers and lakes. It reaches a length of around eight feet and around five hundred pounds. There are several other sturgeons scattered throughout the nation. All of the sturgeons are good eating and their eggs, called roe, make the best caviar around.

Next is the American paddlefish. It is found living in rivers or in lakes and has a unique paddle for its snout. It has a shark-like shape and will grow to around 150 pounds. It is highly valued as food fish and is very tasty whether fresh or smoked. Its roe are almost as good as that of the sturgeon—some say better.

The Gar family is also common in the US, with many different species, though the alligator gar is perhaps the most recognized. This is a real monster, reaching fourteen feet in length and weighing nearly one thousand pounds. It is covered was exceedingly

sharp plates, like scales, and, though it is not known for its flavor, it is large enough to provide a lot of meat. Some of my friends in Louisiana tell me it can be smoked with good results.

The bowfin is also common and is a large-mouthed, sharp-toothed, and elongated fish that is all attitude. It's a powerful predator and feeds mostly on other fish. It is not generally known as a food fish but is around two feet long and weighs eight to twelve pounds.

SHADS

The most common shad in the US is the American shad and hickory shad. Both have been transplanted to the West. They are saltwater fish, but they can be caught in abundance in freshwater on their spawning runs. They are oily but good tasting. The real treasure is the shad roe, which can be fried in clarified butter or bacon fat.

PIKE FAMILY

This family includes a number of fish, but the two best known are the northern pike and the muskellunge. The northern has been transplanted widely, but the muskellunge is still much in its original range. However, it has a hybrid known as the tiger muskie that has been transplanted in almost every state with a lake large. All three are big fish, reaching a length of more than four feet and a weight of eighty pounds. The flesh on each of these is firm and quite palatable when handled properly. Like sucker, these fish have a line of small white bones running the length of each fillet. You need to remove the fillet, lay it down on your cutting board, skin side down. Then, by feeling down the middle of the fillet, you can locate the line of bones. With your fillet knife you can then cut along the bone lines, and you should end up with a triangular-shaped piece of meat. You then add that piece into your stew.

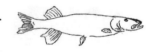

There are numerous other freshwater fish that

could be of use. Don't pass up any other fish just because it wasn't covered here.

There are a sizable number of us that live near the East Coast, the Gulf states, or the West Coast. I will now cover the saltwater species that you might be able to harvest in these areas.

STRIPED BASS

This is a big favorite in New England, down the coast of Maine to New York. I call it the striper. It is the most popular fish in the Northeast. It averages three to twenty pounds and gets as large as eighty pounds. The striper requires freshwater rivers to spawn. They are, at one time or another, available to fisherman from sandy beaches and rocky shores. They respond well to a number of baits, including sea worms, clams, menhadens, eels, herring, mackerel, and anchovies. They also respond well to a large range of saltwater lures.

BLUEFISH

This is the second most sought after fish in the Northeast. It averages from one to fifteen pounds, with a record weight of over thirty pounds. It's not as good at the table, but when prepared right, it's good enough. Its flesh is a little soft and strong tasting. This fish has a mouth full of razor-sharp teeth, and I'm a bit embarrassed to admit I learned about these teeth the hard way. Holding a fish by its lower jaw is so automatic for me that on catching my very first bluefish, I did that without thinking. When I finally got my thumb out of its mouth, I had a line of holes clear through my thumbnail. Blues are voracious and feed like super piranhas on steroids.

WEAKFISH

This fish is the last of the "big three." The average weight is one to six pounds, with a record weight of nearly twenty pounds. But what it lacks in size, it

makes up in flavor. They're mostly a schooling fish, so when you manage to catch one, you know there are others nearby.

COD FAMILY

From time to time, several members of the cod family will venture close enough to shore to be available to someone fishing there. These include the cod itself and the pollock, haddock, and hake. They all have firm, mild flavored white flesh. Many members of the family are large, averaging twenty to forty pounds. They are much sought after by shore fisherman whenever they come near enough to reach.

FLATFISH

This is a popular group of fish. This group includes the Atlantic halibut, the Greenland halibut, and the flounder family—including both the founder and fluke. They are all among the best eating fish in the sea, and they are recognizable by their shape. These fish spend their adult life on their side. One side is white and becomes the belly; the other side is dark and is somewhat camouflage. Some of the larger flatfish prefer deep water, but you can generally catch smaller varieties near shore.

ATLANTIC MACKEREL

This fish is also known as the Boston mackerel. It ranges all up the Atlantic Coast into colder water. Typically, it weighs only a pound or two. Even the record weight for this fish is slightly less than three pounds. It is an oily fish but is well liked by most. It is a very popular inshore fish. It responds better to moving baits or flashing lures.

SCUP

This is a smaller fish that averages a weight of a half-pound to three pounds. The record weight for this fish was nearly five pounds. It's an important inshore

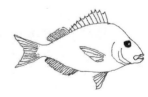

fish. They are not big, but they taste good. They are generally caught using small pieces of clam or squid.

TAUTOG

This fish is also known as black fish. They average twelve pounds. The record was twenty-five pounds. They are a member of the Wrasse family and are excellent eating. They have a close cousin called the cunner fish, which is smaller. While Tautog are commonly caught in deep water, they come into the shore to spawn in the spring, when they are harvested by shore fisherman.

ATLANTIC CROAKER

The record weight for this fish was nearly four pounds. They are a member of the drum family and are good eating. They are the staple fish in Chesapeake Bay. They will take most small baits, but clams are the most often used, and while they are great at the table, they need to be used right away.

NORTHERN KINGFISH

This fish averages around two pounds, with the record at sixteen pounds. Although this is a reasonably small fish, it is a great tasting fish. They have a short barb under their chin. They are very important to the shore fisherman because they favor shallow water and will take most small baits, but they tend to favor sand fleas. They're very strong fighters for their size. In some areas, they are known as whiting.

SPOTS

This is a small panfish, usually under a pound, but they are much sought after because of their abundance and their flavor. They will be found almost exclusively in shallow water, so they offer an excellent advantage to the shore fisherman.

SNOOK

There are a number of different subspecies of snook. Size is generally the main thing that divides each subspecies. However, in general, they will average from one to twenty pounds, with the record weight being over fifty pounds. They are generally found in shallow water and favor mangrove areas or covered areas like docks or bridges. Snook are one of the finest eating fish in the sea. They can be generally identified by a dark black strip down the lateral line into the tail area. Take care when handling the fish near the gills; the area is sharp and can make serious cuts.

DOLPHIN FISH

The word *fish* is used to separate this species from the mammal of the same name. In Mexico, it is known by the name dorado, and in Hawaii, it is known as Mahi-Mahi. They average around thirty pounds, but the record weight is eighty-eight pounds. Without a doubt, this is one of the most beautiful fish in the sea and one of the best-tasting fish as well. This fish is generally known as an open-water fish, but good success can be had catching them from the shore too.

MACKEREL

The most well-known mackerel is the king mackerel. The really big ones are not common near shore, but you can catch fish in the 125-pound range. They come in large schools and can be found in large numbers. The family also includes Spanish mackerel, which averages one to five pounds.

JACK

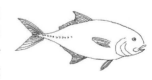

The Jack family includes more than one hundred members. The crevalle jack is the most common and averages two to twenty pounds, with the record nearly fifty-eight pounds. It is probably the hardest-fighting fish found in shallow water. It is extremely aggressive and is very hard to wear down. Its food value, however, is only fair. The Florida pompano will only average twelve pounds, but what it lacks in size it makes up in flavor. Sand fleas are by far the best baits for the small fish. The other good-eating jack is the permit. It is a much larger jack, averaging from five to twenty-five pounds, with the record fifty-six pounds. It is almost as good tasting as the Pompano. It likes to feed on the flat and favors small crabs and shrimp as bait.

DRUMS

The most common drum is the red drum, also known as redfish, which averages three to fifty pounds. This is a big fish that comes into the shallows enough to make it available to shore fisherman.  They have always been regarded as good eating, but when the rage in Louisiana for blackfish started, they were almost fished out. Their numbers are now coming back because fishery managers have lowered the commercial limits and began raising them in hatcheries. The other popular drum is called the black drum. It averages from five to fifty pounds, with a record being well over one hundred pounds. This fish is very good eating.

TRIPLETAIL

I was fishing with a local one day, and he saw the fish floating out near a marker bully. I thought it was just a dry leaf, but he assured me it was a fish, and that he was sure I would want to catch it. I rigged up a bobber from a wine cork with a twelve-inch drop line, to which I added a number-four circle hook. I dug out the smallest live shrimp from our bait bucket, and let her fly. I was just about to reel in and recast when I saw the fish disappear. My bobber then disappeared, and I was into a fight. It normally runs from two to ten pounds. Mine was somewhere around a pound and a half.

WHITING

Fishing for whiting is often best in done canals coming in from the sea. These fish tend to be on the small side, from one to three pounds. They show up in the local markets, usually in five-pound boxes. They sell out fast because they are inexpensive and good eating.

TENPOUNDERS

They are also known as ladyfish. I never understood how it got its name; I've never seen one anywhere near ten pounds, and they sure don't act like a lady. It's not great eating, it has too many bones, and it is as slimy as an eel. But it's a lot of fun to catch. It grows up to three feet long and is very abundant in some places. While its table qualities aren't the best, it is readily available and will help stave off hunger in a survival situation.

RAYS AND SKATES

Let me start with the shovelnose guitarfish. This critter will grow up to five feet long. It feeds on the sandy bottom, so your bait needs to be on the bottom as well. Unlike rays and other skates, these have a long fishlike body. The fat, heavy tail is edible and should not be wasted. Like the rest of the family, it has no bones, which makes it easy to dress and easy to eat. These are often caught during the nighttime hours. In addition, rays and other skates provide edible meat only in their wings. They can be cut free from the body into scallop-like circles, and cooked as you would a scallop.

CALIFORNIA HALIBUT

Of all the fish available to the in shore fisherman in southern California, this would have to be the prize. It's a big fish, growing up to seventy or more pounds. Although this is generally thought of as a deep-water

fish, I have had good success catching plenty from the shore. My experience tells me the big fish are more commonly caught in deep water, but ones in the ten-pound range are common in more shallow waters. This halibut can be identified and separated from the huge Pacific halibut by size. The Pacific halibut can reach over five hundred pounds and lives further to the north. In southern California, the California halibut is king. The flatfish family will also contain many other fish, including other halibuts, flounders, and sand dabs, that are all very good eating.

CORBINA

That is *corbina* with a *b*. It is also at times referred to as California whiting. The corbina is not a big fish, but it is a wonderful-tasting fish that requires skill to catch. It can be caught year-round, but spring, summer, and fall are best times to try for it. It is found on sandy bottoms and generally close to shore. It grows to around twenty inches long and will take sand bugs above anything else. During the day, if you are wearing a good pair of polarized sunglasses, you can generally see the small schools approaching. They are not easy to sneak up on, so you must be careful.

CORVINA

That is *corvina* with a *v*. It is part of the large croaker family, whose largest member is the white sea bass. In addition, there are at least three other corvinas that fishermen look for. They are close cousins to the weakfish and spotted sea bass from the Atlantic, though without spots. They are all excellent eating. The white is the largest, growing up to eighty pounds. The other members of the family are not necessarily little fish either. They range from one to thirty pounds. The larger white sea bass tends to be found in deeper water, but it even it will come in close enough to shore for fishermen to catch as it matures.

CROAKERS

While related to the croakers above, these fish are not the larger members of the family, but they are more available to the inshore fishermen. They are good eating and are always available. The main three are the yellow, the Tommy, and the Spotfin Croaker. For bait, use mussels, clams, shrimp, and sand bugs.

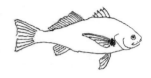

MARKET PERCH

I'm not here talking about one fish but rather several different fish that are called perch at the market. These are often not related, but are all referred to in that manner. The one I chose to represent in my illustration is rightly named California sargo. Some of the others would include the halfmoon fish, zebrafish, and the opaline. They are smaller in size, but good tasting, and—most important to shore fisherman—they feed in and around rocky areas near the shore.

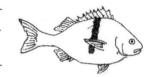

MULLET

This is the most frustrating fish of all because it will not take regular baits. It feeds by sucking in sand and then removing the edibles, letting the sand out of its gills. To make things worse, these fish have very small mouths. They tend to feed in shallow water at the shoreline. It is very frustrating to see a large school of fish swimming near you but refusing to bite. Is there any hope? Yes. The secret is very small dough bait. My recipe is simple: crush fresh shrimp or crab meat into a paste. Then add ripe banana. When it's a good paste, add flour to make it firm enough to stay on the hook. I like to put ten to twelve droppers on one mainline to increase the odds. Spread out your line where you have seen them regularly sit back and wait. Their taste is good enough to be worth the effort. Because they have small mouths, one should use very small hooks.

BARRACUDA
(SPECIFICALLY, THE CALIFORNIA BARRACUDA.)

It is generally much smaller than the Atlantic variety and much less vicious. The really big logs (as they are called) tend to stay out in deeper water, but I have caught a lot of two- to five-pounders near the shore, especially near the rocks. They are slimy to handle but are great eating.

MONTEREY SIERRA

This is the smaller cousin of the more southern sierra. But what it gives up in size, it more than makes up for in flavor. These fish grow to be ten to thirty inches in length. They are considered one of the world's most edible fish.

SHARK

Most sharks are edible, but many, especially the larger ones, have a fairly strong ammonia-like smell to their meat. This is urea in their blood, which helps them deal with saltwater. If you can keep the meat cold for a couple of days, this bad taste will disappear. There are four sharks that seem to get rid of the urea fast enough that you will be able to cook them as soon as you have them. The first is the thresher. It often shows up in the fish market as swordfish. While it is not common near shore, it does come in close to shore in the mouths of rivers. The three other readily edible sharks are the gray, smooth, and hound shark. All three are exceptionally fine food.

HERRING, SARDINE, ANCHOVY, AND SILVERSIDE

These small fish are generally thought of as bait, but a quick look around in your local super market should tell you otherwise. All three are good eating and are eaten by many of us in one form or another.

ROCKFISH

Generally, the rockfish in southern waters tend to stay deeper than the shore fisherman can reach. However, up in the north, a great many rockfish like being in close to shore. This would include the copper rockfish, grass, calico, black and tan, gopher, China, blue, red stripe, smallmouth, and temblor. In California, these rockfish live close to shore and are thereby available to shore fisherman. They're all good eating and abundant. They tend to range from one to thirty pounds. They have a kind of a bass-like appearance.

GREENLING FAMILY

We'll start with the twenty-one-inch-long kelp greenling. The males look so different from the females that they are thought by many fisherman to be two different species! They are excellent food fish and good looking as well. The greenlings are easy to distinguish from other fish because they have five lateral lines. As the name implies, this fish likes to hang out in the kelp. The next most common greenling is the white spot greenling. It is a little bigger, growing up to two feet. Bigger still is the rock greenling. It will grow up to twenty-six inches long. It is usually taken in shallow water over rocky shorelines.

LINGCOD

This fish is a real prize in northern waters. It can grow up to five feet long and is esteemed food fish. It is a shallow water fish, which makes it important to the survival fisherman in the north. Its flesh may have a greenish cast to it when cleaned, but it will turn white when cooked. Its liver is high in vitamin K, so be careful not to overeat it.

CABEZON

This fish is a prize, usually weighing around twenty-five pounds. It is considered by many to be the

best food fish of all. The flesh is a beautiful blue-green, but when cooked, it turns white. Do not eat the fish's row, because it's toxic. It feeds on the bottom and will take everything normally used as bait.

PACIFIC HALIBUT AND OTHER FLATFISH

I know I have already covered the California halibut, but in the north, you will catch the Pacific—a fish that, at the very least, needs to be mentioned. To many Alaskans, it is the king of the sea because of its great size. There are too many flats in the northern waters to mention them all. The one I have caught the most often is the starry flounder. This is only a medium-sized flatfish, and it generally weighs one to three pounds. It is caught year-round.

TOMMY COD

It doesn't grow very large, but it is easy to catch once you locate the school. They are very good eating, but must be kept cool and damp.

PACIFIC COD FAMILY

Most of the members of the cods are represented in the northern pacific.

This is not a comprehensive guide to all freshwater and saltwater fish available. That subject could fill several books! When you are fishing for survival, remember that nothing should be wasted, so put all of your fish heads, backbones, and strips of bones into a pot and boil them for soup stock. This should be the foundation for every chowder soup you make. When the stock is ready, strain the stock through a wire sieve lined with cheesecloth to remove the solids and bones.

When you find yourself in a survival situation, your efforts in harvesting food will likely be rewarded by focusing on fish before big game. It is for this reason I have spent so much time on the fish.

THE DANGERS OF FISHING

You will need to consider how things will have changed and how that will affect how you fish, what you fish for, and most important, what the new level of risk will be. Everything might hang on trying to harvest a meal without giving away your presence. No matter where you are, you will have to be wary and assume everyone in that area is either pillaging and hunting or is, at best, someone territorial who doesn't want you fishing near them. How will that affect the way you fish? Forget about fishing from a boat; being out on the water will make you way too visible. If you are a dry fly enthusiast, you may want to give that up as well. You are forced to make too much movement and even wade out far enough to make yourself visible to others. Even fishing with artificial lures may cause you to move around more than what might be safe. Whatever method you choose to approach the challenge of harvesting fish, you will need to think about not making yourself too obvious. My approach would be to try to find a place to fish where I could stay in some kind of cover, avoiding open shorelines. I would try to focus on bait fishing using spinning gear or bait casting rigs. I would likely choose a bottom rig with a sliding sinker to get my bait out far enough from the shore to give me good odds. I would use a bobber rig wherever conditions would allow my bait to stay out in good water. It will be up to you to decide the level of risk you want to take. If you find you have been lucky enough to be in an area where you are pretty much alone, you may be able to fish a little more out in the open. You should, however, be as cautious as possible, because you can never be sure when the enemy may show up.

FISHING GEAR

What will you do if you find yourself without the tackle you are used to? It is possible you could find yourself without any fishing poles, reels, or even line; what will you do then?

Starting from the bottom up, let's look at what is most important: the fishing line. We could talk about bare-hand fishing, but that is a technique that changes each time, with every attempt. So if we start with fishing as a prepared approach, it brings us back to some kind of fishing line. That could be something as simple as strips of cloth cut from your clothing or string from other things found along the way. There are many things you could use in a pinch, but this is an area I suggest you make some plans and preparations for. The minimum should be some kind of fishing kit. This could include something as simple as fishing line, hooks, and sinkers, all in a small enough package that could be carried on your person.

Your pole and reel could be something as simple as a tin can or a bottle. To make this method work, tie one end on so it will not be pulled off. Wind the rest of the line onto the can or bottle, and attach a good rig to the other end. Once you have set up your rig with a sliding sinker and a baited hook, swing several feet of the baited end over your head and throw it in the direction you want it to go, while at the same time, pointing your fishing reel (the can or bottle bottom) directly at the spot. The line will peel off in the same way it would from a spinning reel, and with a little practice, you should be able to cast your baited hook as far out as you would with a spinning rig. The ideal of course would be to have at least good-quality pole and a plentiful supply of lines and bait set up. A handy fishing kit is always a good idea.

15 ODDS, ENDS, AND SEAFOOD

"When running a marine survival course, there was something I always told my students: if they would take the time to really learn all they could about the seaside environment and its inhabitants, they would never go hungry while living near the sea."

–Larry Mullins

With the possible exception of some areas that have been killed off by pollution, I have never been at a seaside place where I could not find plenty to eat. Learning to locate and prepare members of the clam family alone would make someone starvation-proof. Members of this family have evolved to thrive in every kind of marine environment. They are everywhere. When you learn the needs and requirements of each separate species, you need only to learn the techniques for harvesting the species in front of you.

For some, eating shellfish and other similar foods may be influenced by our food prejudices. When food is scarce, unless you make a serious effort to get over food prejudices, you will end up dead. During my first week in Panama, I went to town with a guy who had been there for years. He took me to lunch at his favorite restaurant. He ordered iguana; I ordered chicken. While we were eating, he handed me a leg from his iguana and told me to eat it. I was repulsed at the thought of eating a large lizard, but I tried it to humor him. To this day, iguana is one of my favorite foods. Think of all the years I would have gone without if I hadn't agreed to try.

CLAM FAMILY

There are too many clams to cover with a single book, so let's start with the clams popular on the Atlantic Coast. I will begin with the clam that seems to be the most popular and the most confusing—the quahog. It is also known as the hard-shelled clam, the little neck, and the cherry stone. Much of the confusion has to do with size. When small, it is called a cherry stone. When it reaches medium-size, it is referred to as a little neck. And when it is fully mature and at its largest, it is called the quahog. When small, it is generally eaten raw or with a good butter sauce. When it reaches the medium-size it's more commonly steamed, and when it's large, it's a lot tougher and is best used in clam chowder. Because this clam as a very short neck or siphon, it cannot dig in very deep, making it easy to harvest. You can use a clam rake, or your feet will do (using your bare feet is called "treading out" clams). The quahog prefers a mixture of mud and sand. Surf clams are also common. The flavor of surf clams is good. I like to open the shell to catch the nectar and put them right into a container with the sand still clinging to the muscles. To keep the full flavor, never wash the abductor muscles, because the flavor will go down the drain. As with all clams, you find the clam bed by looking for an abundance of dead shells on the beach.

Perhaps the biggest and best clam is the razor clam. There are several different species of razor clams, but all species dig into sand fast. You can forget about using a clam rake or your feet to find them. A method perfected in Alaska, called a clam tube, is arguably the best approach to capturing the fast-digging razor clams. A clam tube is generally around two and one-half to three feet long by four inches wide and is sealed at the top with a small hole that you can cover with your finger. It should be wide open at the bottom, and the top should have a handle for you to pull it up. Steel or PVC does a good job. When you spot a hole made by the razor clam, you sneak quietly up and jam the tube down over the hole as deep as you can. The hole in the top lets out the air as you push the tube down. When you hit bottom, quickly put your thumb over the hole and pull up rapidly. The sand and the clam will be caught in the tube and come to the surface. You must move quickly to find the clam before it digs back in to the sand. It takes less than a minute for it to dig out of sight. This approach will likely work well with any razor clam, either in the Atlantic or the Pacific. The whole razor clam family is the best of the best in flavor.

The horse clam is longer than most other clams, and only half of the siphon is usually outside of the shell. Much of the shell is under the sand, so grab the top of the neck and hang on. They are great eating, but should be prepared using the neck separate from the body.

There are several other clams I would like to cover, and many of these are small bodied.

You might ask, "When does a critter get small enough for you to leave it alone?" Remember the rule: if it provides nutrition, it should not be passed over. Rarely is it too small to not be considered. The bean clam, or (as it's known in the East) the coquina, is one such food species. These small clams are meant to provide nectar and don't provide much meat. To prepare them, cover them with water and then boil them fifteen to twenty minutes (many clams will fall out of their shell, and they become part of your meal). Then throw away the shells and enjoy the nectar, which can be used as a hot drink, soup, or chowder. Combined with something else, they will make an excellent meal. Clams thrive on the open sea, but they don't like big waves crashing on them. So they tend to find areas where there are protected, such as behind a sandbar or behind a rocky point. Once you have located them, you can harvest enough for a meal in a very short period of time.

My harvesting tool is a box that was used to hold six-packs of soda pop; I removed the bottom and nailed some small wire mesh hardware cloth to the bottom. I put folding legs on it so I could stand it up in the harvest area. I then put a shovel full of sand in the box and pour bucket of water over them to wash the sand out, leaving a handful of clams. Continue this process until you have a good meal. A side reward is that you can harvest sand bugs, which are also edible.

There are more clams than are covered here, but if you learn the clams covered here well, you will never go hungry.

COCKLE

The true cockle is heart-shaped, and when viewed from the side looks like the perfect Valentine heart. They have a siphon so small they can hardly bury themselves. At low tide, you can walk along and pick them up by the bucketful. There are a few on the Atlantic coast, but the biggest selection is on the Pacific coast. The old song lyrics "in Dublin's fair city, the cry of mussels and cockle, alive oh" bring us to the next thing about the cockle: they are very good eating and, as you likely guessed from those lyrics, are very popular in Europe (especially England and Wales).

MUSSEL

The most common mussel is the blue mussel. It is one of the most abundant, delicious, and easily procured seafoods and can be found on both coasts. The mussel is a veritable storehouse of vitamins and minerals, and it contains some of the most perfect proteins in nature. It was introduced into California well before the turn of the century and has since spread up and down the coast. The mussel prefers to attach itself to rocks and pilings, always facing into the waves.

All you need to do is wait for the tide to start out, then pick mussels until you have enough for a meal. They grow in very thick clusters, and a square yard of rocks should yield enough for a whole family. The adult is about three inches long and is blue-black on the outside. The native California mussel looks much the same. There are several other varieties of mussel, and all are edible. There is one I taught my students to harvest that goes in mud and in the mangroves in estuaries. (I gave it the informal name California mangrove muscle.) It is very good and abundant.

PEN SHELL

This is a large creature that comes in a triangular shell anywhere from five to twelve inches long. The color is somewhat amber. Pen shells avoid both pure sand and pure mud, so look for them in a mixture of both sediment types. They have no siphon, so they must keep themselves at the surface. They are not very difficult to find after you learn what to look for. They prefer warmer waters like Florida and the Gulf states and are quite abundant in Mexico and in southern California. The best part is the large, heavy abductor muscle or scallop, which may be the sweetest, tastiest thing to ever come from the sea. I'm not suggesting you should throw away the rest of the innards, because they are just as good as most other bivalves.

OYSTER

The most common of the oysters are the blue points of Virginia. These are the largest commercial oysters and are some of the tastiest. These oysters do not thrive in the concentrated brine of the open sea, and they tend to seek out the more brackish waters of bays, estuaries, tidal creeks, and even freshwater streams that have diluted waters of the sea. In some places, you can simply pick them up at low tide. Others are deep enough to require an oyster tong.

They were transplanted to the Pacific before the turn of the century and can be harvested on both coasts.

The southern mangrove estuaries contain the coon oysters. These are a lot smaller but are just as tasty. Here you will find them attached to the roots of the mangrove trees.

The best-tasting oyster of all is the wonderful Olympia oyster on the West Coast. Although it is smaller than the blue point, appreciative gourmets are willing to pay more for this oyster. Tidewater farmers do well in cultivating it for commercial sales.

CRAB FAMILY

There are too many separate species to name them all. And while it may seem like I should start with the king crab, I won't because both the king and the tanner crab live in water too deep for the average forager to reach. I will instead cover the crabs that live in shallow water (at least shallow enough for you to harvest from the shore).

Let's begin with the blue crab. It is at home on muddy shores and is usually found in bays, tidal streams, and estuaries. The average adult is about six inches across its back. In other words, it is a medium-size crab. This crab is the number-one crab taken for market. The males can be distinguished from the females by the tales, or key, found on its abdomen. I know of no other crab better eating then the blue. While the blue is an Atlantic crab, it has a very close cousin in the Southern Pacific called the Baja blue. The two blue crabs look identical, except the Baja blue is slightly larger. During many marine survival courses, one of my favorite things was taking students out at night with headlamps and a large landing net. This was best done with the tide out. We would spread out on the sand flats up to our knees and look for crabs. An hour later, we'd have more crabs than we could eat. We would then sit around the fire and talk about how full we were. Another way to catch blue crab is to tie a fresh fish head onto a fishing line, then pitch it out into the water and wait for a bite. When the line starts to move, reel the line slowly in slowly, and then lift the crab over a landing net and it will fall into the net. If you can get a crab trap built, you can also harvest crab from deeper water.

Dungeness crab. The average size of this crab is around six to seven inches, but a big one could be twice that size. The dungeness is common along the length of the West Coast. Like the blue, it favors shallow water.

Rock crab. This crab will measure about four inches across the shell. Similarly, the Jonah crab is common in the north. It is so close to the rock crab in appearance that many people can't tell them apart. On the southern coast of the Atlantic, we find the stone crab: it is also similar in appearance to the rock and Jonah crabs. Most of the meat is in their claws, so fishermen often just pull off the claws and release the crab to grow a new set. The lady crab is smaller and not so popular with fishermen, but in a survival situation, you can't afford to be fussy.

Pacific rock crab. This West Coast crab is slightly larger than the red and has enormous claws that yield enough meat to equal the red and the dungeness. Some gourmets maintain that the flavor of this crab is the best on the West Coast.

Red crab. This is a West Coast crab, and it is usually found near rocky shores. They will grow to be about five inches, and while smaller than the dungeness, its large claws provide so much meat that it will equal the dungeness in the amount of meat.

SHRIMP FAMILY

Shrimp are the number-one treasure to come from the sea. The problem, however, is that they are difficult for the seaside forager to harvest. In general, shrimp are caught by commercial shrimp boats using large shrimp nets.

There was a time when a magazine editor asked me if I would do an article on a family having to survive in a seaside environment. I took my family, which at that time consisted of my wife and five children, ranging from age four to sixteen. We set up our survival camp at a place we call Hector's point, which was a small mountain range sticking out into the Sea of Cortez. The whole range was owned by the Nenninger family. The oldest brother, Hector, was my best friend in Mexico. The whole place was very isolated. The only one who lived there was Hector. The time set by the editor was to be two weeks. We could use only things we could find or catch from the shore. I was very true to the rules, and we ate only things that we were able to catch.

Three days before the end of the set time, though, I let Hector talk me into cheating a bit. He had part ownership in a shrimp trawler and wanted me to go out one whole night on the trawler and see how shrimp were harvested. I took my sixteen-year-old son with me, and together with Hector, we boated out to the trawler before sundown. The trawler worked only at night. That's when shrimp are caught. Great, huge nets are placed out and then pulled along the bottom for hours at a time before it is brought up. The catch is dumped onto the deck. Then the whole crew works to separate the shrimp from the bi-catch. The market shrimp are then iced down in the holding tank while the bi-catch (everything else not shrimp) was shoveled back into the sea. Because shrimpers are after shrimp, everything else was considered trash. The only market shrimp was the Cortez red leg shrimp. In Europe, this shrimp would have been called a prawn. I doubt there is a better tasting shrimp than the red leg. Because my son and I worked all night, the crew insisted we take a large wicker basket full of shrimp as payment.

We realized we had no ice in camp to keep the shrimp from spoiling, so we ate shrimp every couple of hours for two days until the shrimp began to spoil. In those two days, we almost got tired of eating shrimp. We used the spoiled shrimp as bait and thereby turned them into fish for the table. About then, Hector told me how he harvested shrimp wading in the estuary; I had him take me out and teach me how to do it. The secret was in using a net. We used a throw net, which I favor because one person can operate it, but we also had a two-man minnow net that worked just fine. I would carry the throw net over my shoulder looking for my target in the estuary, which was essentially any slight ripple on the surface.

MANTIS SHRIMP

Sometimes referred to as squilla in some places. Despite the use of the name shrimp, it's not closely related to shrimp at all. But it is still tasty, and it can (almost) be caught by hand. While it doesn't have a claw, it has a specialized piece at the end where each claw should be; this appendage is fierce-looking, like a weapon with sharp points to pierce the flesh of its prey. It is folded up like the arms on a praying mantis, hence its name. The Mantis shrimp live in burrows in sand or mud. They are anywhere from six inches to ten inches long. My first experience with this critter was on the Mexican shrimp trawler; every net seem to have a bucket full of them. I had been assigned as the boat's cook, and I decided to try a few.

Collecting them without a shrimp trawler took some thought. As usual, Hector knew how to gather them. His method was to take a twenty-inch piece of fine wire and stick a piece of bait on the end. The foot of a sea snail seemed to work best because it was harder for the mantis to pull off. Then he would make a loop in the end of a length of fish line and spread it around the burrow. The loop was tied with a good sliding knot. Then he slid the baited wire down the hole until he felt the shrimp attack the bait, and then he would slowly pull up the bait, with the mantis still attached, until it popped out far enough to slide the noose around.

LOBSTERS

I will include the two separate species going by the name lobster, even though they are not related.

The first is the northern or Maine lobster. Fishing for them is a huge industry, with millions of pounds sold every year. It would be very hard indeed to find a better tasting food. The main problem to you and I is the need for a commercial license to go after them. However, in times of chaos, a lack of license would not be a problem. But there would still remain a problem in harvesting this lobster, because it will require a boat and a properly made lobster trap. If you are able to work out those problems, it will be worth the effort during the hard times.

Let's move to the spiny rock lobster. It can be found on both coasts, but in the Atlantic it will be found south of North Carolina. It is more abundant than most people think, since it generally feeds at night. It's a big lobster, and I personally like its flavor more than the Maine lobster. It's the only lobster on the West Coast, ranging from Santa Barbara to South America. It has a carapace that is covered with sharp spines, hence the name. Unlike the Maine lobster, it has no claws. Most of my experience with the spiny has been in Mexico or Panama, but I'm sure the information I have gained will be the same throughout its range. Perhaps the easiest approach employed by my students was simply to wade into the water where the

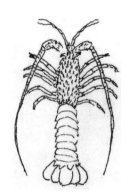

bottom was rocky in the early morning hours with nothing but a landing net. At that time, the lobsters were leaving the shallows and could be gathered with the net. We could also wait until the middle of the day, when we would snorkel the area, looking for a cave or rock overhang. You will need a heavy pair of gloves to avoid getting stuck by the many sharp spines on the lobster. You could also use a fishing pole, and on your line tie a sharp treble hook. You then tie another hook about ten to twelve inches up from the bottom hook. You will bait the top hook, and when the lobster goes for the bait, you swing the whole rig toward the shore. The bottom hook will snag into the body of the lobster and he should come clattering to your feet. In Panama, we would simply go out onto the coral table at night, when the tide was low, and pick them up in the tide pools.

SEA SNAIL FAMILY

Like the clam family, when you have learned to recognize all of the edible sea snails, you probably couldn't starve. In fact you could live on the snails alone. They are that abundant! Again, I plan to include a lot of creatures that might not be very closely related. I will include everything that lives in a snail-like shell.

THE COMMON PERIWINKLE

This is one of the most popular edible snails sold in England today. When the English came to America, they couldn't find periwinkles on our shores. But by 1850, the same species they had enjoyed in England started appearing on our shores. Whether they were purposely or accidentally introduced, no one knows. We did, however, have a native southern species that is quite similar in appearance and flavor. These pose different kind of problem to remove from their shell. Instead of opening the shell, you must remove these by pulling them out. The trick to this is to reduce their size so they can be twisted out. You can accomplish that by boiling them in water with a handful of salt added. The purpose of the salt is to shrink and toughened the meat until it can be "unscrewed" out of its shell. You can also cook them in oil until the operculum falls off. You then stick them with a toothpick and unscrew them from their shell.

DOG WHELK

This is also called the purple snail, so named for the purple dye popular in Greece and Rome. As a source of food, it takes more effort to gather these than it does to gather the periwinkle, and you can't remove them the same way as above. They are tough and require some method of tenderizing them. The best would be a food chopper. They can also be pounded. Once they are made tender, they can be prepared in small patties or in chowder. The situation changes, however, when we move up to larger whelks. Though large whelks are very abundant in Europe, they rarely show up in our markets (likely because of American food prejudices, though I think the flavor is great). To eat them, you must hammer the foot so you can begin your preparation. Another way is to sprinkle tenderizer on the foot, as you would an abalone. You do this after cutting the foot into half-inch steaks. All whelks are delicious and easy to harvest.

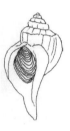

MOON SNAIL

Moon snails are anywhere from four-and-one-half to five inches. They can be found on both coasts and are generally found in shallow water, making them readily available to the survival forager. When they are not on the move, they like to cover themselves with a layer of sand. You prepare and cook their large foot like you would any other sea snail previously mentioned.

ABALONE

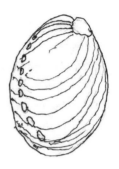

When you look at the abalone, you might think it isn't a snail. However, if you look closely at the top of it shell, you will see the evidence of the whorl pattern of the snail, indicating that it is in fact a member of the snail family. It is a serious vegetarian and eats only algae. This has the advantage of not causing you to worry about shellfish poisoning and, unlike so many other snails, it has not been the subject of food prejudices. It was gathered by the coastal Indians for thousands of years, but with the coming of the Europeans, the pressure on it became too great, and its numbers declined. They are currently found in low numbers and nowhere near the abundance of the past. These low-population numbers have become serious enough that the size limits and harvest numbers are now strictly enforced. In a period of chaos, though, these laws would mean little. There are quite a few varieties, even on the West Coast, and the largest is the red abalone. Pressure has pushed it into deeper water, so snorkeling equipment will be necessary. In addition, there is the green abalone. The black abalone and the pink abalone are more common in the north. Others include the Pinto abalone, the white abalone, and the flat abalone. Harvesting them requires prying them off of the rocks where they are attached. For this, you will need an iron to slip under the shell and pry it off. The only edible part is the foot, but it's large enough to make a meal. Slice the foot into three-quarter inch steaks, then place each steak between cheesecloth, and pound them tender with a meat mallet. You will never eat anything better.

EDIBLE LIMPET

Largest of the limpets is the owl limpet. It is about three inches in length and is shaped like a squat Indian tepee. Again, we are rarely interested in anything but its foot, but the foot is sweet and tasty.

To be honest, I have never cooked limpets because I enjoy eating them right from the sea.

CHITON

Another critter I have had the same problem with is the chiton. Just foraging along, eating as I go, is the way I like them. Most chitons are fairly small, but the best one is the giant sea cradle. It is a giant and big enough to truly be worth the trouble, because it can grow up to thirteen inches in length. Again, it's just the large foot you'll want to eat.

SEA URCHIN

My guess is you have never thought of this prickly critter as food, but you would be wrong. All of my marine survival students learned to enjoy eating them. The part of the urchin we use for food is its roe. It is delicate, delicious, and the very best caviar. The way to go about harvesting roe is to cut or break off the bottom. Be careful because this critter is nothing but a ball of sharp spines, and in some urchins, the spines can be mildly poisonous. The biggest and best is the giant urchin. It ranges from Alaska to Mexico. The next most popular is the California purple urchin. While not quite as big as the giant, it is still large. The best time of year to harvest the roe is in late summer to midwinter. The roe is best eaten raw, spread on crackers or crispy bread. The color of the row is bright orange.

EDIBLE SEAWEED

The most popular variety of edible seaweed would be laver, then rockweed, which is popular in clambakes. After that comes dulse, sea lettuce, and Irish moss. Most people like seaweed better after it has been dried. Seaweed is a good source of nutrition, and once you develop a taste for it, it is very enjoyable to eat.

OCTOPUS AND SQUID

Squid will be hard for the average forager to catch, but not impossible. Octopus are very good eating and are much easier for the forager to find. To do this, simply turn over stones at low tide. When you find one, grab it quickly, then bite hard over the eyes; this paralyzes the octopus and will cause it to die sooner. Then you can pop it in your carrying sack and move on, looking for the next octopus.

OTHER SMALL SEA CREATURES

There is a small shrimp you may think too small to eat called krill. This small shrimp, however, is almost the sole food for many whales, penguins, and other large creatures. I try to eat a bit of krill each day because it is extremely good heart food. In addition, there are many very small fish under an inch to four inches long that you may not see as food because there is no way of preparing them. They are still good nutrition and not to be passed up. The idea is to fry them whole or dry them whole and then eat them. I have already talked about such things as sand bugs, sand fleas, and other small crustaceans. They may make very good bait and food. Earlier, I referred to the sand lancet. This is an interesting, primitive, edible fish. It is only a few inches long, and it looks transparent. It has no fish head, but it does have a sucker mouth with cilia around the inside edge. Lancets can be obtained by washing them from the sand through a screen container, leaving the creature unable to dive back into the sand. I like to add a few in a stir-fry.

I have not talked about the many small snails and small crabs, all of which are edible but are difficult to remove meat from. They can still be boiled so that their nutrients create a kind of soup.

If you're on a dock or at the end of rocky point, shine a light into the water. The light will attract a multitude of small creatures. I have made a wonderful soup by simply scooping them up with a wire sieve and cooking them along with other things gathered throughout the day.

DRY LAND FORAGING

There are many things I would normally cover in this section that we already covered in the chapters on animal food and fish chapters. I will now try to be careful to cover only those things not already covered. If you are living away from the water, whether it be fresh- or saltwater, what could you find to eat? Let's remind ourselves what the situation is that we're dealing with: absolutely no food is available except for whatever you have managed to store or what you can forage.

INSECTS

"No way!" you say. "Now you really lost me." All I am asking is that you hear me out. Away from the water, what potential food do you know that is more abundant than insects? "But are they edible?" Yes. Nearly all are edible. The US army survival guide tells us they are "the most abundant life-form on earth. . . . Insects provide 65 to 80 percent protein compared to 20 percent for beef."[1]

Let's start with your food storage itself. Unless you have just recently purchased it, or you took special care in sealing it up, your storage could already contain insects. Things like wheat or other grains often contain weevils, flour moths, or mealworms.

If you open a storage container and find something like that, what would you do? We're talking about hard times. You may not feel like throwing away the whole container. But what options do you have? You could use the grain for planting, but it would be a long time until you will be able to harvest something to eat. You can try to separate the grain from the weevils, and then eat the grain. However, the weevils have tainted it already. Or you could grind it all together and put it in a dough that you bake into bread. What I do is separate the weevils from the whole thing. Then I grind the wheat and make bread and grind up the weevils and put the ground up paste in my stew. The flavor is great. The protein is there to keep me strong and active. The best part? There are no insect parts, just flavor and nutrition.

Mortar and pestle. In this process, you take whatever insect you have obtained and put it in a pan containing salted water and bring it to a boil (most survival foods should be boiled first to kill anything in them). Next, put them into your mortar and begin to grind them up into a paste. This process removes anything that might be objectionable. You will now have a paste that contains almost all of the nutritional value with none of the hard outer shells. As a fine paste, it will literally disappear in anything you add it to. It will work great in soups, stews, and casseroles.

Cheesecloth method. For this method, cheesecloth is the best, but any porous material will do. As always, the first thing is to catch the insects. Members of the grasshopper family work best for this method. In the grasshopper family, there is one member that is not edible called the lumber. It is a very large grasshopper with a red color on its shell. It is mildly poisonous. The rest of the family is edible. Once you have harvested the insects, place them in the cheesecloth and tie them into a ball. The size of the cheesecloth required will depend on how many insects you have captured. Then you simply drop the ball of insects into the stew you already have boiling. Continue cooking for a couple more hours, then pick the ball of insect bodies out of the stew and throw them away. The stew now contains all of the protein, vitamins, and nutrients but none of the body parts.

The "break 'em up" method. My students, during our thirty-day survival experiences, were not all to be talked into the "insect for supper" experience. But a good percent were coaxed into it if I agreed to remove the wings and legs first. I've never been sure why that should make such a difference, but it did. The way we usually handled the preparation was to remove the legs and wings, then put the remaining insect on a thin green stick and roast it over the fire.

The "oh, just eat 'em" method. I know a number of survival experts who say once a person has learned to enjoy insects as a food source, that person will never starve in the wilderness. Now, I'm not sure that works with the family trying to live off their food supply, but we all know they would always be able to find something to eat. Maybe I ought to share with you my first experience in this area.

I was a young man working for the US Forest Service in a program designed to eradicate pine bark beetles. The first day, I was assigned to work with a couple of old veterans. As we walked along spraying trees, I noticed the older men would rip off a large slab of bark on an infected tree before spraying. I begin to watch closer to see what it was they were doing.

One of them, seeing my inquisitive attention, offered to show me: they were picking out the beetle larva. Whenever they got all the larvae available on that tree, they would move on. But now their left hand was full of larvae, and all day long, they would pop those little bugs in their mouths. These larvae looked like pine nuts, both in size and shape and taste. By the second day, they had shamed me into trying one. After that, I quit carrying my lunch bucket, and because I was never hungry at noon, I was able to catch a nap instead. Learning to eat insects is the best way I know to make someone starvation resistant. Not all insects are that easy to gather or eat, but it is worth your time and energy to find those you like and will use. Starvation is not a great deal more pleasant than most people would expect. The body begins fairly quickly to become autocannibalistic, or in other words, it begins to eat itself. At first it's not a big deal. The leftover glucose from food is devoured first. The fats follow. The next step is when the proteins begin to be consumed from the muscles. This is followed by the tendons. The individual loses strength, which in turn begins to compromise the individual's ability to work at saving him- or herself and others. So, the number-one rule is, "No reasonable nourishment should be overlooked." Also of prime importance, "While your strength is near its maximum, do not pass up any promising source of sustenance." When food becomes no longer available through the means normally available to you, all the rules change! Your former taste prejudices are a luxury you can no longer afford. If survival is your goal, and it surely will be, these prejudices can no longer be allowed.

In addition to saltwater snails, it is useful to know that our dry-land snails are also edible. Larger snails are the best because more meat can be gathered with less effort. Snails should be handled just like their saltwater cousins. Boil them in salted water or oil so you can unscrew them from their shell to eat. Freshwater clams are also good too, but be careful not to harvest them in polluted water. I have gathered a whole handful of great eating clams from a fast, cold stream.

A word about crawdad or crayfish: they are just miniature lobsters. When my family lived in Dutch John, Utah, near the Flaming Gorge Reservoir, my boys and I would put out our crawdad traps on Friday and pull them in Saturday afternoon. Every Saturday night, we dined on large crawdad. We put them into a large pot with some crab boil and cooked them until they turned red. We then spread a large cloth on our kitchen table and dumped the crawdads on the table. We ate until we were stuffed.

While I fully understand you will not be happy to hear this last item on our survival list, I would be remiss to not cover it: earthworms. They are very high in much-needed nutrition. They can be cut open to remove the inside line containing what they have been eating. Either way, to deny yourself the source of nutrition might well be the one thing that could have saved your life.

NOTES

1. *The U.S. Army Survival Manual Department of the Army Field Manual* (Berkeley, CA: Ulysses Press, 2009), 21–76.

16 PRIMITIVE SURVIVAL SKILLS

"You might be in a situation where all you have is what is on your person at the time. Can you go primitive? You will not be able to survive if you cannot readapt to technology that is thousands of years old."

–Unknown

Most of us, when planning for TEOTWAWKI, think we have covered everything. After the SHTF, however, Murphy's Law tells us if anything can go wrong, it will. Learning primitive survival skills will help us overcome most situations—especially the unplanned ones. We may have planned everything out to the letter. We start off toward our bug-out place, where everything is waiting for us that we will need to ensure survival, but then run into some unforeseen wild mess that cuts us off from our best-laid plans. This unforeseen mess may be so extreme that we find ourselves alone in the woods, running for our lives with nothing but the clothes on our backs.

Take this scenario: As a good prepper, you did everything you were told you needed to do. No one said anything about a situation like this. You feel you are prepared; your bug out vehicle is ready; you have it packed with everything you will need while on the road, including several days of food and water. You have enough weapons to defend yourself, and you have previously prepared your bug-out dwelling with everything you will need to see you through a couple of bad years. You decide things have gotten bad enough for you to head to your bug-out place. You and your family climb into your vehicle and start out of the city. The highway is packed but moving along. You make it out of the city, and then, little by little, you leave the suburbs behind. About sundown, you come to a stop behind stalled traffic. Your bug-out vehicle is designed for off-road use, so you begin looking for a way; as you find it, you move out of the jam. You continue inching your way forward until you are stopped. You step out of your vehicle, and as you do, you see a number of armed men near the jammed vehicles around you, and they are killing the vehicle's occupants. The armed men are running toward you too. You turn to the safety of your bug-out vehicle, but just as you do, you become aware your family is out and running toward the woods. You run to catch up with them, and, as you turn to leave, you see that the armed men are killing everyone behind you. You catch up with your family as they enter the woods. You all continue running deeper into the woods, trying to get away from the sound of shooting. Back at the highway is a scene of mass chaos. To try to return to your bug-out vehicle would be asking for death. Your choice is simple—you must go on with what you have. You cannot return to your vehicle, and you cannot reach any of the things you have prepared. Several techniques are provided here to help you survive in the event that you are separated from your stockpile of supplies and gear. These techniques, for the most part, assume that you will have very little on your person and will be living primitively.

Building a fire is one of the most important skills to ensure your survival. Several fire-making tools (some would say "gimmicks") exist to make fire starting easier. These tools include (but aren't limited to) a magnesium stick, fire stick, magnifying glass, cedar puck, weatherproof matches, Tinder Quik, and pairing batteries with steel wool. These are all good items to have and generally work well. However, assuming these tools are not available, basic fire-starting tools and techniques can be obtained from your environment.

MAKING FIRE USING NATURAL MEANS

Flint and steel is perhaps the oldest and most basic fire-starting method. Primitive and historic outdoorsmen, like the old trappers of the western US, would have carried flint and steel kit in their "possibles" bag. The kit would include a good piece of flint (also called chert or chalcedony), agate, quartz, or other similar rock that sparks well, and a piece of high-carbon steel. Be aware

that stainless steel, which is common in many outdoors knives, doesn't spark well. In today's world, the majority of pocketknives have stainless steel blades because they don't rust and they look nice. But many stainless formulas don't produce hot sparks, don't sharpen easily, and don't hold an edge well when they do. The key to getting sparks, then, is the carbon in the steel. Besides a carbon steel pocket or belt knife, other metals that spark well are a piece of old mill file, pipe, or other chunk of construction material. After you have your flint and steel, spend time learning to make good, hot sparks with them. The time to learn this skill is well before you need it in a survival setting. Most people think the spark is made directly into twigs or grass, but that will almost never work. When using a spark to start a fire, you need something you have previously prepared. Again, the historic woodsman shows us the way. He would prepare some cloth, which he usually carried in a tin. (I use a three-inch film can.) The idea is to char the cloth without burning it up. Real linen will work best. Cotton will also work as long as it doesn't contain other chemicals. Put the cloth in the can and set it afire. Then quickly put on the lid to cut off oxygen and wait a few minutes for it to cool. The cloth should be black, but still holding together. It is this carbonized cloth that catches the hot spark. A hot spark is a spark with enough carbon left to ignite the carbon in the cloth. And how do you get a hot spark? Practice, practice, practice.

Fire bundle. A fire bundle, or something I call a fire nest, is also an invaluable tool. Preparation, too, is a critical step to a successful fire. In some ways, making a fire is a little like prepping for surgery. The surgeon has all the tools laid out on a tray in order. The tray should be on a cart so it can be at the surgeon's fingertips. In the middle of

brain surgery is not the time to go looking for some needed tool. Making fire from scratch is a little like that. Before you even think about making a spark, you should have everything laid out at your fingertips. The fire nest should be previously prepared and ready to receive the spark. Fine, dry grass or something else highly volatile should be laid out and on hand. Larger twigs, finger-sized sticks, broom handle-size sticks, and larger branches, in sequential order, should also be available. There is nothing as disheartening as getting a fire started and then losing it while you wander around looking for materials to feed it. Build your materials into a position before starting. As soon as the nest bursts into flame, simply slide it into the tepee of tinder and sticks that was previously prepared.

The fire nest may be the most important part of fire making without a match or other fire starting tool. Whether you use the flint and steel method, bow drill, hand drill or any other approach, you will always start with the fire nest. If you build it properly, it will look just like a bird's nest. In fact, I have used actual bird nests in place of one built with my own hands.

To build a nest, start with the outside and work in. On the outside, you fashion this bird nest shape out of dry, volatile materials. One of the best materials in the West is the bark of the juniper tree. Sagebrush bark and rabbitbrush also work well. In the South, bald cypress, Spanish moss, and palmetto leaves will work. Bark from white cedar, shagbark hickory, or redwood will work well too. I'm sure you get the picture. Any dry and volatile bark that can be stripped away and then shredded makes good nest material. Shred the bark by rubbing and working it back and forth between your palms. This breaks the bark down so it can be molded around into a round nest shape. The bark needs to be woven and tied into itself to hold it in place. Your fire bundle or nest should now be holding itself together. The next step is to repeat the outer layer with another layer, but this layer should be crushed up even finer between your palms, then wound into a circle and worked inside the outer layer. The third layer must be broken down even finer and then carefully worked inside the coarser outer layers. The last layer will be material broken down as fine as you can make it.

Depending on what method you are using, you'll still need this small bed to contain very fine fire material. When using the flint and steel method, use a fine material that could almost catch the spark by itself, such as thistledown or the down from the virgin's bowers. The last piece to go into the nest is a couple of pieces of burnt charred cloth—in the center of the nest where the spark will land. Then, taking flint in one hand and steel in the other, strike sparks into the cloth, watching carefully for any sign of an ember developing from the well-placed spark. As soon as an ember is detected, close your hands around the bundle, raise it up above eye level, and begin to blow. Blow soft and easy at first. This is the most delicate time in the fire making process. If you don't see smoke in the first minutes of blowing, you have lost the ember and need to begin again.

If you see smoke coming out of your fire bundle, continue blowing and, as long as you're getting smoke, don't pause; keep the air blowing in a steady stream. At some point, the bundle will burst into flame, and when it does, quickly place the burning bundle on the bed of dry grass you have prepared as the base for your fire. Then begin to add the matchstick-size twigs slowly until you can see the fire has moved up into them. Once they are burning well, you can begin adding bigger twigs and sticks. Once that is all burning well, you can begin to add larger pieces of wood. Up to this point, you can't lose focus, because losing focus means losing your fire. But as the fire grows, you can stand up and bring larger pieces of wood to the fire. This might seem like a lot of extra work, but failing to make a fire and spending a cold night without one will change your mind.

Take the fire-making process seriously, and give it forethought. Prep the fire right the first time, and you will have success. While primitive skills are simple, they are not easy to learn. In many ways, primitive skills are as much an art as a skill. Fire making is a good example of a primitive art. I have worked with students—often for hours—before they could get fire and build a fire consistently. These were hours filled with failure. Often, when success finally came, the student couldn't say what they did different from the previous failures. Once the skill is acquired, though, continue practicing and working at it until you can repeat fire making with regular success.

Bow drill. When building fire with the bow drill, your fire bundle or fire nest is no different than the one you put together for flint and steel. The big advantage to the bow drill, however, is that you do not need a previously assembled kit. Once you become proficient with the bow drill, you can assemble everything you need for the bow drill from nature. After you learn how to make cordage, you can make your own string.

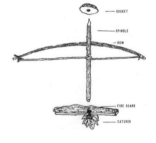

To make fire with a bow drill, it is important to get your materials properly assembled. The best place to begin with is the fireboard. This is the base foundation of the system. It should be about a half-inch thick, and it needs to be relatively flat so it won't roll or move. You will hold it down with your foot. Unless you have a good knife, this will be one of the hardest things to do because you will very likely need to split, cut, or whittle the board into the proper shape. Flat pieces of sagebrush are common at the bottom of the plant; this will need to be broken off at both ends. Cottonwood works very well but will also need some shaping. Stay away from hardwoods because the fireboard, to work right, should be very easy to drill. You are trying to produce a lot of fine powder material from the board, and it is in this fine powder that the ember to produce the fire is found. For the same reason, the spindle should also be softwood.

After the fireboard is shaped and ready to go, it is time to prepare the hole and notch where the fire magic happens. Start by making a depression a half inch in from the edge. This will be the hole. It should be just deep enough to hold the spindle in place as it is spun (it will get deeper as you work). Once you have the depression finished, you then need to cut "V" notch from the edge into the depression. As you work the spindle, the fine powder, mentioned above, will build up as you work. As you work the bow and spindle, smoke will also start coming from the hole. Don't stop too early when you see smoke. You will need to get at least a teaspoon of powder before you can create an ember. The powder contains small burnt particles of wood, which the friction heat has turned into a kind of charcoal. When a certain temperature is reached, tiny particles begin to spread a small cluster of embers and they begin to produce smoke.

You need to put a "catcher" under the notch. The catcher can be a sturdy leaf or something similar that will transport the ember and the remaining powder from the fireboard bundle. Transporting the ember must be done carefully—the worst thing is to lose the ember while transporting it to the fire bundle. Ideally, the spindle should be a quarters of an inch in diameter. I have used rabbit brush and sagebrush, but my all-time favorite is yucca. Sharpen the top and make the bottom blunt. For the bow, find a branch with a curve or that is itself bowed. It is also helpful if you find one with a small fork on one end. This is useful to adjust the tension in the string. The socket needs to be something that can be made smooth inside the hole where the spindle fits. You will not be carrying it with you on a daily basis. I do, however, usually carry my socket with me as a lightweight soapstone rock. The soapstone is also slick inside allowing the spindle to turn smoothly. The last component is the string, or cord. A good choice for your cord is your bootlaces. Strips from your pants can also work in a pinch. Natural cordage can be made (as detailed bellow) as well.

Hand drill. The hand-drill kit is similar in concept to the bow drill, but you won't need a bow, socket, string. The spindle will also be longer and thinner. I have one survival colleague who is very good at the hand drill, but I have never gotten good with it. I have managed to make it work for me several times when doing demonstrations for students, yet I can't say I have developed a real skill at it. Consequently, this is not a method I am quick to recommend to beginners. The basic principle is the same as with the bow drill. Instead of using a socket, though, the spindle is twirled between the palms of your hands. My colleague who excels at the hand drill tells me the secret is in the hands and working the spindle up and down between your palms. The spindle should also be about twenty inches long and thinner than the bow spindle.

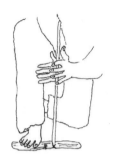

THE NOT-SO-PRIMITIVE METHODS

This is what is often called the optics approach, or using the sun to help you make fire with a magnifying glass, bottom of a quart bottle, or other thick piece of glass. The glass needs to be clear. A piece of clear ice can also be shaped with your warm hand into the proper shape. The main issue here is something to focus the sun's rays into a pinpoint of light on some flammable material. Another very good method is a magnesium fire stick. They are relatively inexpensive, so buy a half dozen. Put one on your key ring so you will be more likely to have it when you need. The others could be scattered around in places where you'll have one when you need it. The ferrocerium rod and steel striker also work well. I have a few, and they work great. If I were to count on a match, it would have to be waterproof. Then there is using a battery and steel wool, but what is the likelihood you will have both on the day you need them? Learn to use the primitive methods covered in this chapter, however, and you will be more confident. You'll be able to get a fire going when you really need it.

WATER

There is simply nothing more important than water. When TEOTWAWKI happens, people will worry most about what they're going to eat, but the more important issue should be getting water to drink. Ideally, an individual can last more than a month without food but can last only a few days without water. Being out in the woods without water means you have to first find natural water or produce water.

Building an evaporation still will often work to produce water. To make a still requires a sheet of clear plastic, but once in place, the plastic and something to capture moisture in is all you need. Dig a hole, place your container at the bottom, and suspend the plastic above the container. Anchor the plastic along the edges with rocks or sand and place a rock in the center to weight it toward the container. Sunlight will condense water onto the bottom of the plastic, which will then drip into your container. You can also place the plastic sheet over a tree branch facing the sun and seal it. Water will condenses in the plastic and drip to the bottom.

To find water, start by looking for obvious sources—a spring, stream, pond, lake, or reservoir. These sources are often referred to as open water or water open to the surface. To find water, look for trees and plants that require it (such as cottonwood) or green areas in the bottom washes, canyons, and

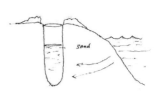

riverbeds. Look for canyons, large or small, because canyons often contain springs, seeps, or other sources of water. Check out canyon bottoms for running water. If the ground is muddy or damp, try digging to reach standing water. If you dig down around fifteen inches and nothing seems to be changing, try somewhere else.

However, finding a source of open water, like a small stream, does not mean your water concerns are over. There are a number of things you need to be aware of before you can use the water you found. If the water is open to the environment, it will likely not be pure and will contain a number of waterborne bacteria harmful to humans, such as giardia (which is found in any open water where animals drink). Besides giardia, there are a multitude of other harmful bacteria inhabiting our waterways and water sources. After you drink contaminated water, you'll likely get a case of screaming diarrhea every ten minutes, which is bad enough by itself, but it will also induce severe dehydration. Without medical attention to treat the diarrhea, you could die within days. You simply cannot risk illness as a problem.

After finding water, you must kill the bugs to make the water safe and drinkable. There are a number of things designed to help you with that, from filters to chemical additives. But in a survival scenario, you will not likely have these tools in your pockets. Boiling the water might be the only technique available to you. You must have fire before you can have water, again emphasizing the importance of being able to produce fire.

To bring that water to a boil will require something you can put over the fire, such as a container to bring the water to a boil. That container will work best if it is metal, but a canvas bag or a plastic container will work too in a pinch. You can heat rocks in your fire, and when they are hot enough, put them into your nonmetal container until, rock by rock, you bring the water to a boil. The key to survival is your ability to adapt. If your source is a small stream or a spring, technically, you should be dealing with normal bacteria. To be safe, boil the water for about ten minutes. But just bringing the water to a boil will kill most bacteria.

But how much water should you be drinking? Try a gallon a day. Most of us don't drink anywhere near that amount during our average day, but in a survival scenario, you should drink a lot more water than you think. You will be working harder than you ever worked before just to live. Having to spend all of your time out of doors will be draining water from your body much faster than your normal activities would.

SHELTER

You will need to retain as much of the water in your body as possible. Being able to get in out of the sun will go a long way to helping with that. Wind is as bad as the sun in robbing your body of its moisture. A good shelter is the best solution to these problems and will help keep you warm and out of the rain and snow. So once you have fire and water issues resolved, the next item of consideration is shelter.

The first serious decision you will need to make is where to build your shelter. Television survivalists give the impression that the only consideration is finding a level spot, but it's a little more complicated than that. For example, building under the tallest tree in the area is like asking to be hit by lightning. Building next to a stream in the bottom of the canyon might get you washed away in a flash flood. Additionally, cold settles to the bottom of canyons, so building up away from the bottom may increase the temperature ten or twenty degrees. Moving up from the bottom will be warmer and safer. Try to find a spot close to the materials you are planning to use in constructing your shelter. Choose a location not too far away from your water source but far enough away that your bodily wastes won't contaminate your water. When you think you have found a good spot to build, stop and look around: are there dead trees close enough to fall on you in a strong wind? Are there hazards that you might stumble into during the night? Things like poison oak, loose rocks, or deadfall? The next big decision to make will be the kind of the shelter to build.

Lean-to. One of the most basic and easy-to-construct shelters is the lean-to. It's not considered the warmest structure, but it could still save your life. Begin by finding two standing saplings, alive or dead, and lash a horizontal branch between them to serve as the backbone of the structure. Place (or lash in place) vertical branches, then cover the branches with twigs, leaves, and other materials to create a wind barrier and cover. Tarps and blankets can also be used to form the structure. Wrap the edges around to create wind shields at the side. Root balls (or root walls) formed by overturned trees are also a convenient starting point and will save construction time. You can then build your fire toward the front of the structure. Surround your fire with rocks, which will contain the fire and help store and radiate heat.

Desert small tepee. You may find yourself in a desert environment with no trees or high brush to work with. Don't be discouraged. It is possible to build adequate single-man dwellings out of sagebrush or rabbitbrush. You begin by digging out a depression to serve as the center of the dwelling. Next, find four good saplings still rooted with the right space between them. These four will be the anchor for the dwelling. You can find stick them together like a small tepee, leaving the tops together to hold the weight. You then continue to stack the limbs, but now use limbs that are thicker. When you get to the point where you're not seeing

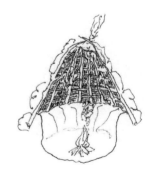

too much light from inside your structure, switch to adding grass bundles. Keep checking from the inside for holes, and place grass or other vegetation where you can still see light. Even in heavy storms, you will stay dry for days in the structure.

The basket wickiup. This is a survivalist's go-to shelter when it is needed in a hurry. To begin, find a spot with at least four tall sapling trees separated by an area for the shelter. Pull the saplings together at the top and tie them in the middle with vines, leaving a space underneath large enough for at least two people. Next, cut down plenty of other trees about the same length, bury the bottoms of each in the soil, and then pull them together and tie them at the top. The gap between each stick should be no more than ten inches when finished. Cut more sticks, but they don't need to be as thick. Weave them in and out between the upright sticks until there is a layer of sticks moving up toward the top. Now all that is left to do is to attach thatch. This can be long grass, pine boughs, palmetto, or Spanish moss. Any large leaf or slab of tree bark will do. My favorite, however, is longleaf plants like cattail. Bulrush will also work with a little more effort. The plan is to attach your thatch all around the first layer of horizontal sticks, and then move up to the next row and so on until you reach the top. You will need to leave an area attached right at the top for a smoke hole. If you could turn this structure upside down, it would look like a woven basket. Now move your fire into the center and make yourself comfortable.

The rustic tepee. In many ways, this structure type is not so different from the basket wickiup. In this one, you will stack larger, longer polls with no binding. You must begin by gathering plenty of long poles. They should be at least twenty feet long. You'll need to gather at least a dozen such poles, and at least one

of the tallest poles needs a forked end. Start by setting the first three poles together at the top to bind them together. Then lay in other poles, one by one at the top. You'll need to keep an even space between them. When they are all in place, you began to weave in or tie-in horizontal sticks all the way around the structure. Then move up a foot and repeat the process. These horizontal sticks are what you tie your thatch to. This tepee type of shelter works best in a coniferous forest because that's where you will find the type of long, straight poles needed to make it work. Again, leave an opening at the top for the smoke to pour out, and you're ready to face the elements.

The long house. The eastern Indians often used this type of structure because it works best in the kind of environment common to the eastern US. This is also a good structure for larger numbers of people. Instead of the basket shape, you will be making something more like a Quonset hut. You'll need to find a place where you have several tall young trees in lines. Pull two of them over toward each other and tie them at the top. You can cut other poles and plant them in the ground facing each other and tie them together. When you finish, they should be no further than twelve inches apart. Next, tie in a series of horizontal sticks (this will hold your thatch), but leave space in the front as a doorway. Sheets of birch bark can be used as thatch. Once finished, build fire in the center and move in.

Caves. Caves are especially useful structures in the western deserts and are a convenient, nature-made shelter to find when you don't have a lot of time to build a good structure. Nearly any cave or overhang can be turned into a reasonable dwelling by simply adding some stones or deadwood. You need to leave enough of an opening door and then build your fire in front of the door.

TOOLS FOR GATHERING FOOD

There are several primitive tools that will make gathering vegetable and animal foods easier. Several are outlined below.

Throwing stick. I don't think you will find anything simpler to make than a wooden throwing stick. Many Native American tribes used a throwing stick as one of their most basic food-gathering tools—and many, like the Hopi, still do. Interestingly, they prefer the wood of rabbitbush plants or greasewood plants for their sticks. Both plants produce a heavier stick. If harvested at the bottom of the plant, they produce a straight stick with a knob on the end. This knob improves the deadliness of the throwing stick by adding weight at the impact end. The stick should be about two feet long and be at least as big around as a broom handle. One does not have to be extremely accurate when throwing the stick because it covers a two-foot wide swath as it travels. If the rabbit is anywhere within that swath, it's rabbit stew.

Fish spear. This is one of easiest primitive tools to make, and is also one of the most effective. To make one, take a long hardwood shaft and split one end into four prongs. Tie off the split about six to eight inches below the end so the shaft won't split further. Then slide a wood block down to the tied-off spot. This will force the four separate prongs away from each other. Sharpen the ends of each prong, and you are ready to go fishing.

Slingshot. No weapon has been around longer. David proved it could bring down someone as large as Goliath. With enough practice, you should be able to hit a target as small as a sparrow. The best part is all you will need is two long strings and a little leather from the tongue of your boot. The ammunition is free—rocks—lying on the ground all around

you. Swing the sling around your head and let the rock fly; with practice, your aim and lethality will improve.

Bola. A bola consists of three leather strings of cord. The best material to use for the strings is leather, especially rawhide. Each cord is about three feet long, with a rock about the size of a billiard ball encased in a rawhide pouch at the end of each cord. This weapon will bring down most animals you run into, including game-sized animals. To use them, hold one of the balls in your hand and spin the whole set over your head; when you're ready to release, throw the ball in your hand as if you were throwing a baseball. The bola will fly toward its target in a huge, six-foot wide circle. It will not matter where the bola makes contact with the target because it will wrap itself around the target and bring it down.

Bow and arrow. If you are a novice to archery, the bow and arrow is one of the most difficult primitive skills to master. To succeed, keep it small and keep it simple. Start with a bow. You should concentrate on making a quick bow because you will not have the skill or time to build a good bow. The first thing is to find a branch the proper size. If you are an average-sized person, a good length will be about forty-five inches long. Look for hardwood. A piece that is nearly straight but with a slight bend will be best. Smooth and work the wood with a piece of coarse rock. The branch should be dead and well seasoned. You can straighten the wood by wetting it, holding it over your fire, and then shaping it over your knee. Notch both ends to be able to receive the bowstring. The bowstring will be the hardest part of the whole project. You'll need to find some kind of heavy string or cord at least fifty-eight inches long. For arrows, you will need to look for brush with straight stems. Willow is a good choice, as is wild rose. Next, you will need to find something to use for fletching. Of course, feathers work best, and hopefully birds are

available at your location to provide the feathers. Use animal sinew to tie the fletching onto the arrow, as well as to hold the arrowhead in place.

Atlatl. This weapon has been in use long before the bow and arrow. The spear consists of a long, thin shaft about six feet long. The spear is attached to a throwing stick—called the atlatl—that is used to propel the shaft with greater force and distance than is possible by arm alone. The atlatl is about two feet long and two inches wide with a hook on the end to hold the rear of the spear. It has finger loops near the other end. The spear is attached to the atlatl and then thrown. Like an arrow, the spear should be fletched to make it fly truer. The atlatl is relatively easy to make but is not as easy to throw as it sounds and requires practice to use effectively.

Boomerang. The aborigines of Australia still do most of their hunting with a throwing stick called a boomerang. They have developed the technology to a higher state of perfection than any other people. When people think about the Australian boomerang, they think of the self-returning boomerang, but that is incorrect. As far as I can tell, the returning boomerang is basically a toy. The serious hunting boomerang doesn't return; its purpose is to fly straight and fast and hit hard. The hunting boomerang is about a yard long and only slightly curved. One side should be flat and the other side is curved to form an airfoil—like the wing on an airplane. When properly thrown, it spins, and that spin can carry it five hundred to seven hundred feet. As it flies, it cuts a swath three feet wide, giving the hunter a greater chance of hitting his or her target. To make one, look for the heaviest wood you can find. Look for a limb with a curve about six to eight inches from the center. Carve the airfoil shape by rubbing it on sandstone or other rough rocks. When finished, you can quickly develop skill and accuracy with it, and

you will be astonished how successful it can be in procuring fast game.

Traps and snares. Trapping food is arguably the easier and more effective way to gather animal protein. A large number of traps or snares can be set simultaneously, providing higher odds of success than hunting. There are a number of variations of both traps and snares, so to avoid complicating things, we will cover only some of the most basic sets.

By learning the "figure four" trap, you can build an almost unlimited number of different traps. This trap requires only three sticks. You can use this trap under logs or rock slabs. Carve flat notches in each of the support sticks and connect these together to make a support frame for the slab or log. The bait stick is usually carved into a pointed end. The key to making it work under different conditions is the bait stick. Bait is attached to the end of the pointed stick, and when the animal takes the bait, it pulls the setup from under the log or slab, and the animal is crushed. I also like to drive two sticks on either side of the deadfall to make sure it falls straight.

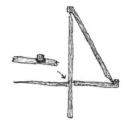

The stick sapling snare is also simple and effective. The sapling snare, like most snares, requires a string or cord. I favor a fishing line called Spectra because it's hard to see, hard for an animal to bite through, and is super strong. To begin, locate an animal trail, look for a sapling that is about three feet from the trail. Strip off any limbs that might affect the action of the sapling, build the snare, and then when everything is finished, pull down the end of the sapling and attach it to the string. At the end of the string, tie a large slipknot and pull to make the loop about the size of the animal you're trying to snare. You can use small twigs on either side to hold the loop open.

MAKING TOOLS FROM STONE

When you are thrown into a survival situation without a knife, you will be nearly overwhelmed at first. Most food items require a cutting edge to process, and you will need one to prepare structures, snares, and other items. But don't give up yet. You can make a primitive cutting tool almost anywhere you find yourself. With training and practice, you can turn stone into a tool you'll be proud of.

Hand axe. The simple approach is to use your eyes rather than tools to make the hand axe. In other words, look until you find a rock the size and shape you feel you will need. Try to find a rock that already has an edge-like shape. Chert, basalt, agate, quartzite, and other similar rock types work best. Then, find a second rock (or hammer stone) of harder material to knock (or flake) off pieces from the edge to sharpen it. The flaking should be conducted along both sides of the axe head to form an angled edge, not unlike the cutting edge of a modern metal axe. Remove as many flakes as needed to form and sharpen the edge. Next, find a spot about midway through your ax head, and, by using your hammer stone, peck out a channel on both sides of your axe head. This is where you will attach your ax handle, so the channels should be deep enough to hold the handle in place but not so deep that it becomes a weak spot where the stone head might break. Next, using the ax head, cut a handle from a hardwood sapling, and then make a split in the center of the top of the handle. Be careful not to split it too far. If you have managed to kill an animal, you can use sinew or tendon from the animal to tie the head into the wooden handle. If fresh animal hide is unavailable, you will have to sacrifice some article of your clothing. Using your ax head, cut the material into thin strips. Then, using one of the strips, wrap the split ends to keep the handle from splitting any further. Slide the ax head into place. Last, force the top of the split together, and tie it off with another strip of belt. Your axe is now ready to use.

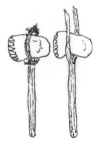

Stone knife. Because we have no way of knowing what kind of rock formations you'll find yourself in, I will try to keep this section simple. Whatever kind of rock you find in the area, look for something that makes flakes, such as fine-grained basalt, chert, obsidian, agate, jasper, or the glass bottom of a big bottle. With your hammer stone, work along the edge, removing flakes from the downside of the rock. Continue working the edge back and forth until it becomes sharpened. If you want to make a spearhead or arrowhead, use the same process. You can also use things like deer antlers as tools for creating these weapons. The base of the antler becomes a hammer for larger work, and the antler point or tine becomes the tools needed for finer pressure work. You should develop skills while working on the cutting tool or knife that you can then apply to the finer work involved in making other tools. Obsidian is the easiest stone to work with when making blades, spearheads, or arrowheads.

Epilogue

This book was never intended as a comprehensive "how to" survival book. The goals of this book have been to help make it clear what kind of things we might experience in an end of the world scenario and to prompt us to think more deeply about what will follow the nation's collapse. It has also been about how to deal with the aftermath of such a fall. It is my fervent hope that you will all have discovered some new ideas, facts, and ways of looking at this upcoming event.

We need to remember that the kind of scenario has never happened before in our nation. The big question is how will people to respond when the problem is that there is no food available anywhere, for any price, and on a nationwide scale. We'll have to face a world where the vast majority of the population will turn to looting, pillaging, and murder to get food. This will not be the case with serious peppers. They have seen the writing on the wall and have made themselves ready. But the preppers are only a small part of the population.

There are some who would say the only way I could prepare for the future is if I could see the future. However, that's not necessarily true. There are a number of ways you can prepare yourself for the things you believe might happen. I don't prepare myself for some precise event. I prepare myself for a more general kind of thing—a general event. I feel I am prepared to deal with any number of things that can go wrong. Things I will need to adjust myself to, if I have any hope of surviving. The argument might come back, saying, what if nothing like that happens? My response would be to say, "Wonderful! What have I lost by preparing? Nothing really."

We have been given instincts to help us anticipate the dangers ahead. I have been dealing with those feelings for many years. For me, those feelings have been so certain that I have spent most of my life preparing myself and helping others prepare themselves to be ready.

At the point in time when SHTF, how will you respond? Your response will be the difference between whether you live or die. Here, I will assume you have made the best preparations you could. You have a plan. You have put away at least a year supply of food and water. You have purchased the weaponry needed and have trained yourself to use it well. If your plan is to bug out, you have done all you could to make that option work. If your plan was to stay put, you have made yourself ready.

But there is still one thing you still need to address—your emotional preparation. To address that preparation, start with this key: "The will to survive can also be stated as the refusal to give up." In any survival situation, nothing is more important than attitude. Your first mission is to stay alive, and the second is to help keep your loved ones alive. So how do you go about preparing to give yourself the best chance to survive?

TRAINING

There is nothing that will give you the confidence you will need like training for anything that might come up. This will require you to learn what you might have to face. This will mean having some kind of idea what will likely happen and how you will handle it. If you find yourself having to face people who are there to take your life and the lives of your loved ones, then you must have made a decision about what you are willing to do before you are being placed in that situation. And if you have decided to face the force with equal force, you will have to have trained yourself in the ability to carry out your side of the confrontation with skill and deadly effectiveness. You will have had to make decisions about what kind of weapons will serve you best. You should have confidence that the other members of your family are trained well enough to do their share in that defense. But in the end, if everything that could go wrong does go wrong, you should still have the confidence that you will still be able to handle it. Simply refuse to give up.

FEAR

Don't pretend that you have no fear. That is only self-delusion and self-denial. Everyone has fear. The goal is not to eliminate the fear but to build confidence in your ability to function despite your fears. Being unexpectedly thrust into a life-and-death situation is not something you can fully prepare yourself for. You will suffer a certain degree of fear. But you can prepare yourself to handle those fears and learn to rule over them. Fear is generally an emotional response to dangerous circumstances. Fear can immobilize us and cause us to become so frightened that we fail to perform activities essential for our survival. You can be sure that when SHTF, it will constitute the most terrifying thing you will have ever experienced. Fear will affect us all, but again, the key will be how fast we can get that fear under control.

ANXIETY

This is a kind of persistent fear. While it is not as the debilitating as outright fear, it can still wear you down. It is much harder to control, but you need to control it as much as you can. In a survival situation, it will help to stay busy performing tasks that are necessary to keep you and your family alive. In other words, stay busy!

ADAPTABILITY

No one can guess how long the hard times will last. Clearly, it would be a mistake to think of this as a short-term issue. If the situation is going to require us to live under these conditions for a long period of time, then we will need to adapt to living conditions similar to the 1800s and not only survive but thrive.

HUMOR

In a book on survival, it would seem to be a strange place to talk about humor, but as it turns out, humor can be a very powerful tool in helping you deal with a high-stress situation. People working in a high-stress situation, like combat medical personnel, law enforcement officers, firemen, and combat troops have learned that if they can't find some way to lighten the stress, they will fall apart.

Let me try to explain this mechanism as I understand it. First, we need to understand how our mind works. It is composed of two main parts: the conscious and the subconscious. We are aware of the conscious part. The subconscious is that part we are not consciously aware of. The subconscious gets its information about the world from the conscious. If the conscious is in panic mode, the subconscious will be worse. However, if the subconscious is receiving lighthearted humor, it takes the message that things must be fine. So it does not go into panic mode, and can thereby still be helpful and creative. You stay reasonably able to think clearly and work out each challenge as it appears. This doesn't necessarily mean you should go through the day laughing and joking. It simply means seeing the humor in the simple things throughout the day. It means keeping your head in a light place rather than a dark and fearful place.

Humans have been able to survive shifts in their environment throughout the centuries. Man has shown the ability to adapt physically and mentally to a changing world. The ability to adapt is what has kept us alive while many other species around us went extinct. This ability kept our forefathers alive, and it can keep us alive if we'll have confidence in our inherent ability to adapt.

About the Author

Larry Mullins learned his survival skills during his years growing up on a ranch and serving in the US army. Those skills came in handy when he directed the BYU survival program for several years. He also taught security skills at LDS Church Headquarters and developed the bodyguard program designed to protect the President of the Church, eventually leading teams that protected President Spencer W. Kimball. After that assignment, he served in various other forms of law enforcement.